Everyday Mathematics®

The University of Chicago School Mathematics Project

STUDENT MATH JOURNAL

VOLUME 1

The University of Chicago School Mathematics Project

Max Bell, Director, *Everyday Mathematics* First Edition; James McBride, Director, *Everyday Mathematics* Second Edition; Andy Isaacs, Director, *Everyday Mathematics* Third, CCSS, and Fourth Editions; Amy Dillard, Associate Director, *Everyday Mathematics* Third Edition; Rachel Malpass McCall, Associate Director, *Everyday Mathematics* CCSS and Fourth Editions; Mary Ellen Dairyko, Associate Director, *Everyday Mathematics* Fourth Edition

Authors
Jean Bell, Max Bell, John Bretzlauf, Mary Ellen Dairyko, Amy Dillard, Robert Hartfield, Andy Isaacs, Kathleen Pitvorec, James McBride, Peter Saecker

Fourth Edition Grade 3 Team Leader
Mary Ellen Dairyko

Writers
Lisa J. Bernstein, Camille Bourisaw, Julie Jacobi, Gina Garza-Kling, Cheryl G. Moran, Amanda Louise Ruch, Dolores Strom

Open Response Team
Catherine R. Kelso, Leader; Amanda Louise Ruch, Andy Carter

Differentiation Team
Ava Belisle-Chatterjee, Leader; Martin Gartzman, Barbara Molina, Anne Sommers

Digital Development Team
Carla Agard-Strickland, Leader; John Benson, Gregory Berns-Leone, Juan Camilo Acevedo

Virtual Learning Community
Meg Schleppenbach Bates, Cheryl G. Moran, Margaret Sharkey

Technical Art
Diana Barrie, Senior Artist; Cherry Inthalangsy

UCSMP Editorial
Lila K. S. Goldstein, Senior Editor; Kristen Pasmore, Molly Potnick, Rachel Jacobs

Field Test Coordination
Denise A. Porter

Field Test Teachers
Eric Bachmann, Lisa Bernstein, Rosemary Brockman, Nina Fontana, Erin Gilmore, Monica Geurin, Meaghan Gorzenski, Deena Heller, Lori Howell, Amy Jacobs, Beth Langlois, Sarah Nowak, Lisa Ringgold, Andrea Simari, Renee Simon, Lisa Winters, Kristi Zondervan

Digital Field Test Teachers
Colleen Girard, Michelle Kutanovski, Gina Cipriani, Retonyar Ringold, Catherine Rollings, Julia Schacht, Christine Molina-Rebecca, Monica Diaz de Leon, Tiffany Barnes, Andrea Bonanno-Lersch, Debra Fields, Kellie Johnson, Elyse D'Andrea, Katie Fielden, Jamie Henry, Jill Parisi, Lauren Wolkhamer, Kenecia Moore, Julie Spaite, Sue White, Damaris Miles, Kelly Fitzgerald

Contributors
John Benson, Jeanne Mills DiDomenico, James Flanders, Lila K. S. Goldstein, Funda Gonulates, Allison M. Greer, Catherine R. Kelso, Lorraine Males, Carole Skalinder, John P. Smith III, Stephanie Whitney, Penny Williams, Judith S. Zawojewski

Center for Elementary Mathematics and Science Education Administration
Martin Gartzman, Executive Director; Meri B. Fohran, Jose J. Fragoso, Jr., Regina Littleton, Laurie K. Thrasher

External Reviewers
The *Everyday Mathematics* authors gratefully acknowledge the work of the many scholars and teachers who reviewed plans for this edition. All decisions regarding the content and pedagogy of *Everyday Mathematics* were made by the authors and do not necessarily reflect the views of those listed below.

Elizabeth Babcock, California Academy of Sciences; Arthur J. Baroody, University of Illinois at Urbana-Champaign and University of Denver; Dawn Berk, University of Delaware; Diane J. Briars, Pittsburgh, Pennsylvania; Kathryn B. Chval, University of Missouri-Columbia; Kathleen Cramer, University of Minnesota; Ethan Danahy, Tufts University; Tom de Boor, Grunwald Associates; Louis V. DiBello, University of Illinois at Chicago; Corey Drake, Michigan State University; David Foster, Silicon Valley Mathematics Initiative; Funda Gönülateş, Michigan State University; M. Kathleen Heid, Pennsylvania State University; Natalie Jakucyn, Glenbrook South High School, Glenview, IL; Richard G. Kron, University of Chicago; Richard Lehrer, Vanderbilt University; Susan C. Levine, University of Chicago; Lorraine M. Males, University of Nebraska-Lincoln; Dr. George Mehler, Temple University and Central Bucks School District, Pennsylvania; Kenny Huy Nguyen, North Carolina State University; Mark Oreglia, University of Chicago; Sandra Overcash, Virginia Beach City Public Schools, Virginia; Raedy M. Ping, University of Chicago; Kevin L. Polk, Aveniros LLC; Sarah R. Powell, University of Texas at Austin; Janine T. Remillard, University of Pennsylvania; John P. Smith III, Michigan State University; Mary Kay Stein, University of Pittsburgh; Dale Truding, Arlington Heights District 25, Arlington Heights, Illinois; Judith S. Zawojewski, Illinois Institute of Technology

Note
Many people have contributed to the creation of *Everyday Mathematics.* Visit http://everydaymath.uchicago.edu/authors/ for biographical sketches of *Everyday Mathematics* 4 staff and copyright pages from earlier editions.

www.everydaymath.com

Send all inquiries to:
McGraw-Hill Education
8787 Orion Place
Columbus, OH 43240

ISBN: 978-0-02-143087-1
MHID: 0-02-143087-X

Printed in the United States of America.

8 9 10 LMN 20 19 18

Unit 3

Contents

Unit 1

Unit 2

Unit 4

Multiplication Facts Strategy Logs

Facts Inventory

Activity Sheets

Welcome to *Third Grade Everyday Mathematics*

DATE

Dear Children,

Welcome to *Third Grade Everyday Mathematics*! This year you will use what you learned in other grades to do more math and become an even better problem solver.

Here are some of the things you will do in *Third Grade Everyday Mathematics:*

- Add and subtract larger numbers.
- Estimate and round numbers.
- Learn multiplication and division facts.
- Solve number stories.
- Measure length, time, mass, and liquid volume.
- Collect, organize, and represent data.
- Model and compare fractions.
- Tell time to the nearest minute and solve problems about time.
- Recognize and measure shapes.

As you solve problems and number stories, you will

- Make sense of problems and solutions.
- Work together with your classmates to solve problems.
- Explain your thinking to your classmates.
- Listen as your classmates explain their thinking to you.
- Keep trying even when problems are hard.
- Think about more than one way to solve problems.
- Use math tools to help solve problems.
- Look for and explain patterns.
- Create rules and shortcuts to help you solve problems.
- Learn to use your *Student Reference Book* and other resources.

Mathematics is all around you. We want you to become better at using mathematics so you can better understand your world.

Sincerely,

The Third-Grade Authors

Finding Differences on a Number Grid

Lesson 1-1

DATE

									0
1	2	3	4	5	6	7	8	9	10
11	12	13	14	15	16	17	18	19	20
21	22	23	24	25	26	27	28	29	30
31	32	33	34	35	36	37	38	39	40
41	42	43	44	45	46	47	48	49	50
51	52	53	54	55	56	57	58	59	60
61	62	63	64	65	66	67	68	69	70
71	72	73	74	75	76	77	78	79	80
81	82	83	84	85	86	87	88	89	90
91	92	93	94	95	96	97	98	99	100
101	102	103	104	105	106	107	108	109	110
111	112	113	114	115	116	117	118	119	120

Use the number grid to help you solve these problems.

1. Which is less, 83 or 73? 73 How much less? 10
2. Which is less, 13 or 34? 13 How much less? 21
3. Which is more, 90 or 55? 90 How much more? 35
4. Which is more, 44 or 52? 52 How much more? 8

Find the **difference** between each pair of numbers.

5. 71 and 92 21
6. 26 and 46 20
7. 30 and 62 32
8. 48 and 84 36
9. 43 and 60 17
10. 88 and 110 22

Looking Up Information

Work with a partner. Use your *Student Reference Book* for Problems 2 and 3.

1 Write your partner's first name. Sriyatsan

Write your partner's last name. Srinivas

2 **a.** Locate and read the essay "Number Grids."

Describe what you did to find the essay.

____

____

____

b. Do Check Your Understanding Problems 1 and 2 on page 92.

Check your answers in the Answer Key.

Problem 1a: ______ Problem 1b: ______ Problem 1c: ______

Problem 2:

a.

		58	59
66	67	68	

b.

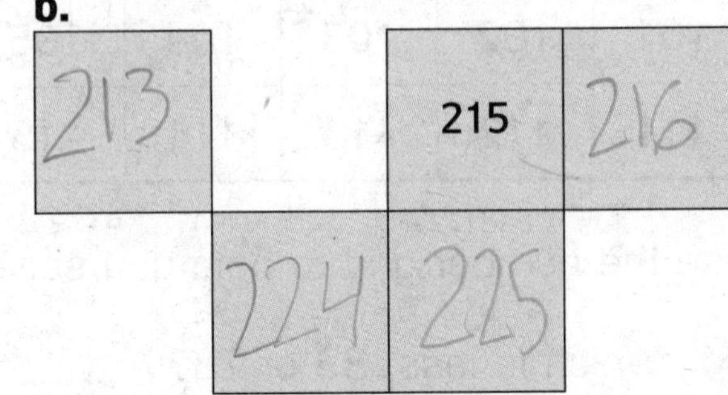

c.

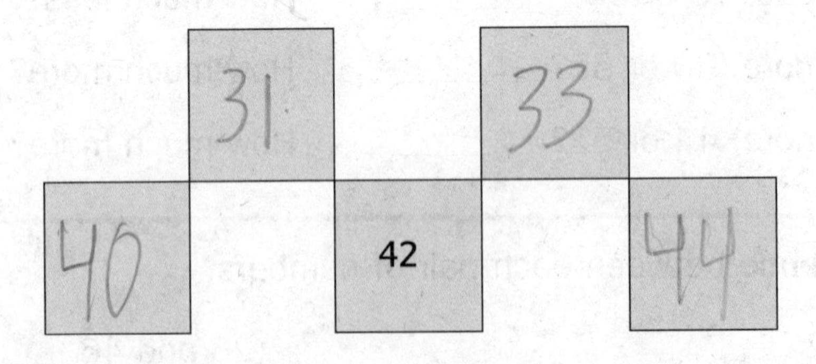

3 Look up and read the rules for *Number-Grid Difference*.

On what page did you find the rules? ______________________________

Using Mathematical Tools

Lesson 1-3

DATE

For Problems 1 and 2, record the times shown on the clocks. For Problem 3, draw the minute and hour hands to show the time.

1 

8:30

2 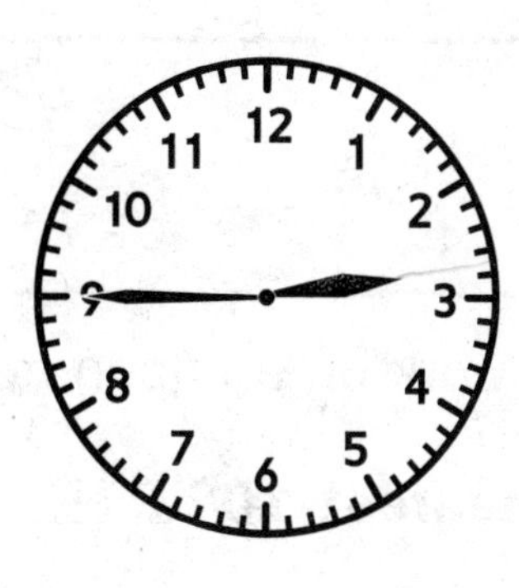

3 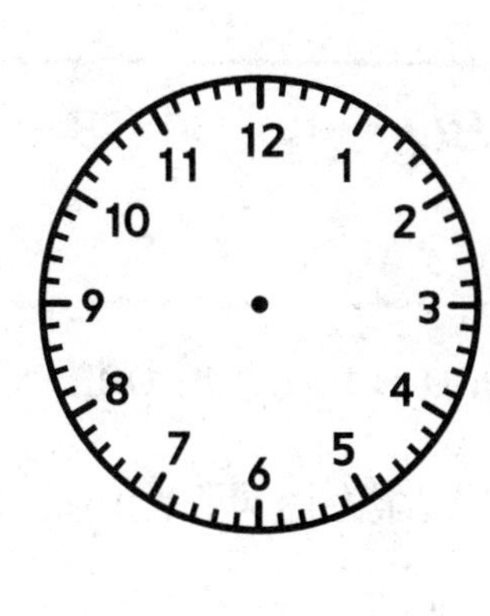

6:10

4 Use your ruler to measure the line segment.

This line segment is about 3 inches long.

5 Draw a line segment 10 centimeters long.

Use your calculator. What keys did you press to make each change?

6 Enter 50. Change to 107.

107-50 =57

7 Enter 94. Change to 30.

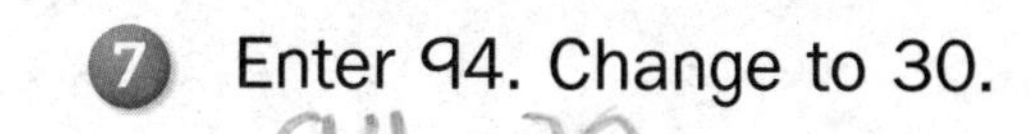

94-30=

8 Use your Pattern-Block Template. Trace two polygons that have exactly 4 sides.

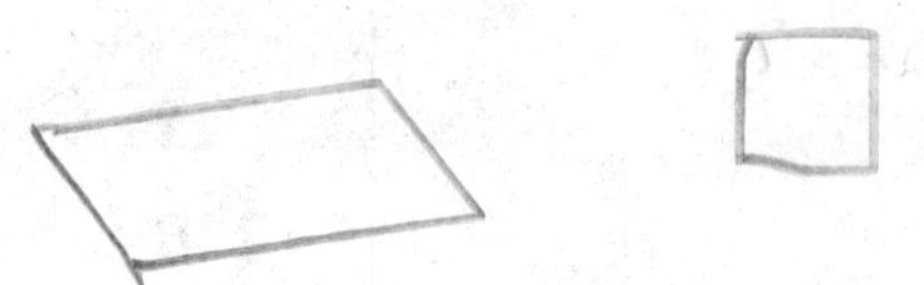

What are polygons that have 4 sides called? (Use your *Student Reference Book* if you need help.) quadihteral

Rounding Numbers

DATE

Example: What is 83 rounded to the **nearest 10**? 80

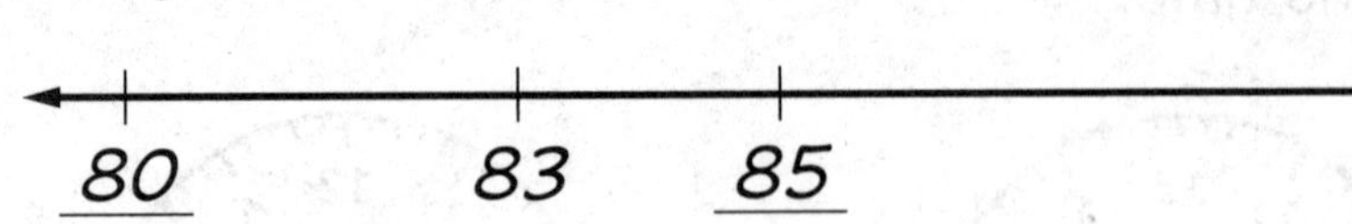

Which two multiples of 10 are closest to 83?

Round each number. Show your work on an open number line.

1. What is 47 rounded to the **nearest 10**? 50

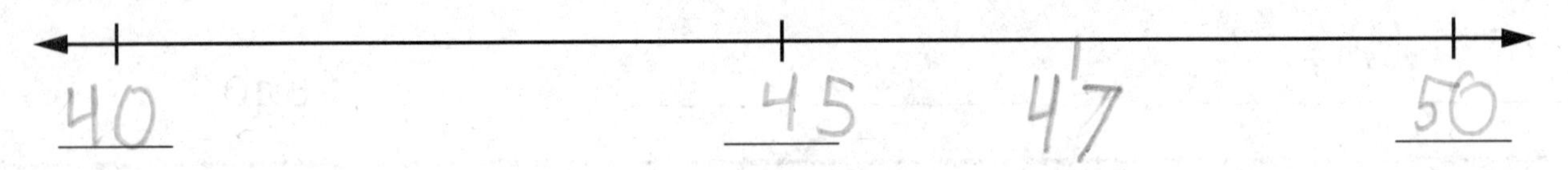

2. What is 72 rounded to the **nearest 10**? ______

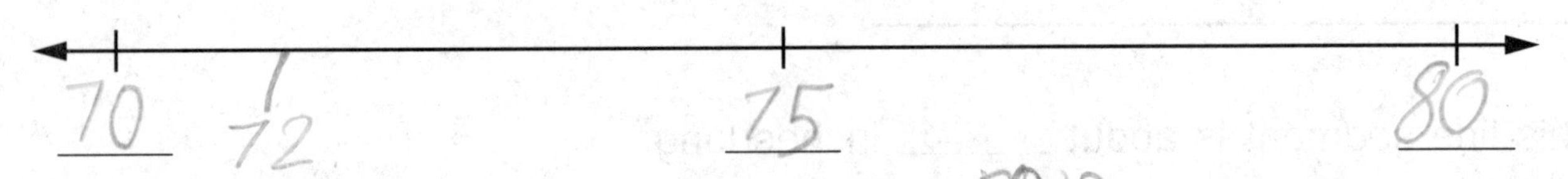

3. What is 234 rounded to the **nearest 100**? 200

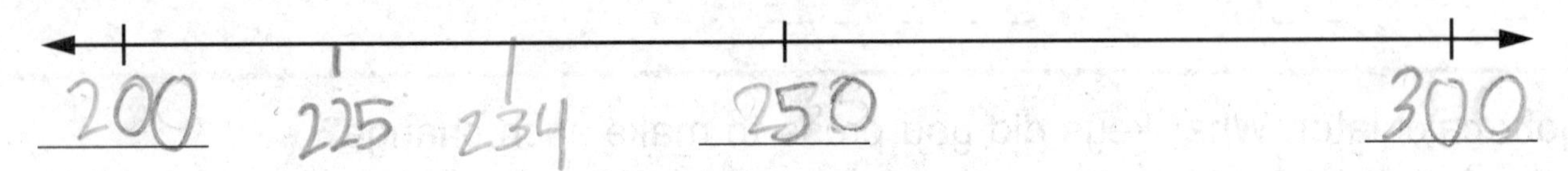

Try This

Round each number. You may sketch number lines to help.

4.

Number	Rounded to the Nearest 10
43	40
97	100
453	450

5.

Number	Rounded to the Nearest 100
348	200
89	100
297	300

0 50 100

Telling Time

Lesson 1-5

DATE

Math Message

Circle the clock that shows the time.

1. Which clock shows 1:30?

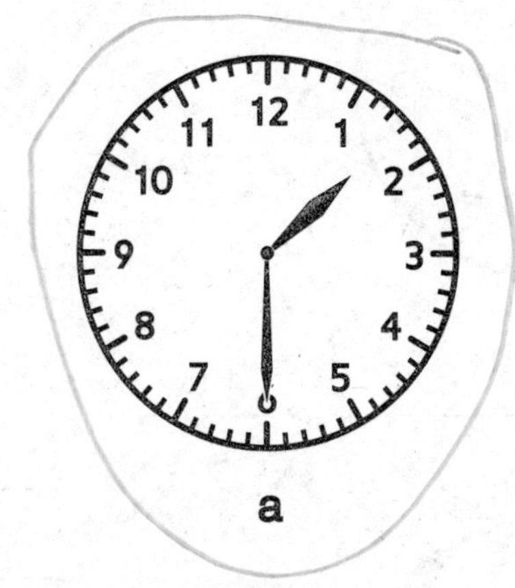

a

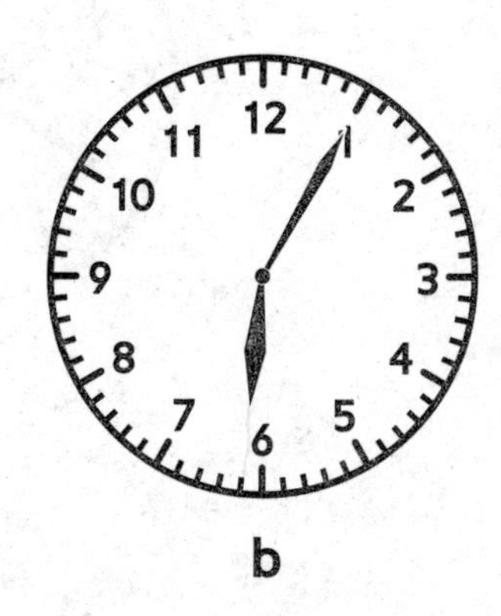

b

2. Which clock shows 2:45?

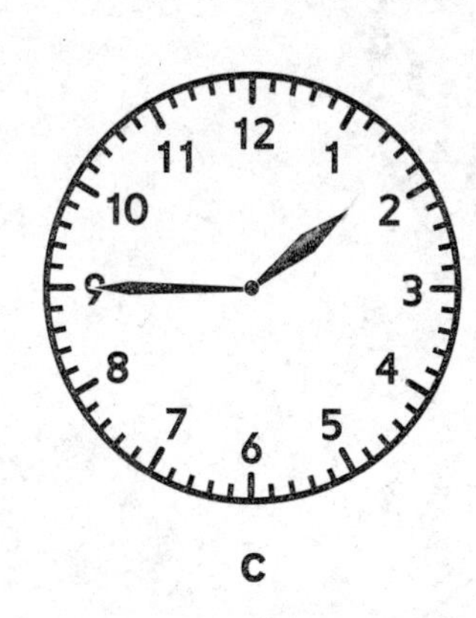

c

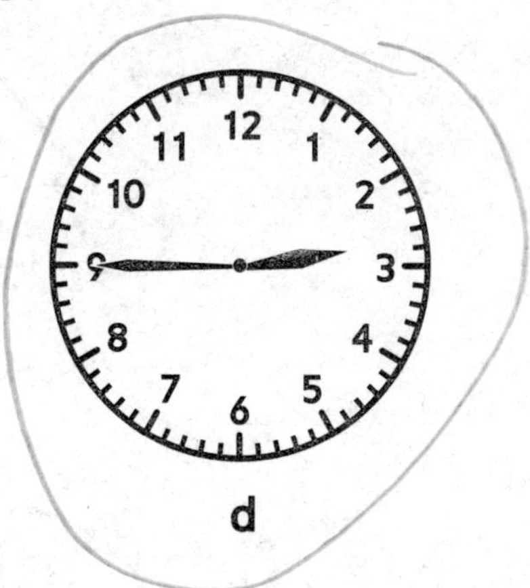

d

3. Which clock shows 4:55?

e

f

Reading Time to the Nearest Minute

Lesson 1-5

DATE TIME

Write the time shown on each clock.

1

2:15

2

2:40

3

3:30

4

3:37

5

10:00

6

10:13

Math Boxes

Lesson 1-5

DATE TIME

1 Write the numbers that are 10 less and 10 more.

10 less		10 more
28	38	48
235	245	255
357	367	377
809	819	829

SRB 89, 98

2 Use your calculator.

Enter	Change to	How?
31	41	+ 10
58	38	−20
146	106	−40
405	455	+50

SRB 293

3 Fill in the unit box. Add.

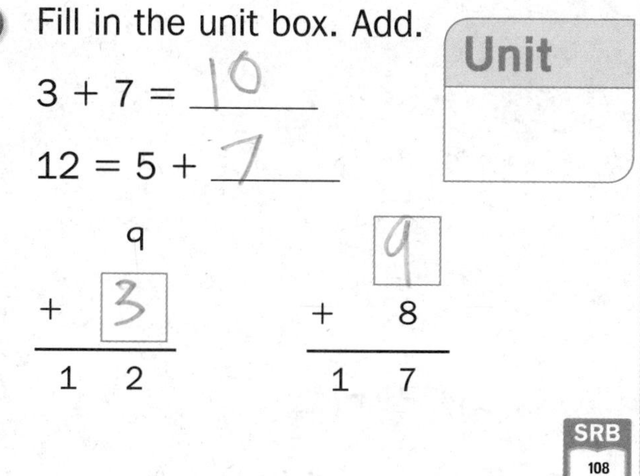

Unit

3 + 7 = 10

12 = 5 + 7

$$\begin{array}{r} 9 \\ + \boxed{3} \\ \hline 12 \end{array} \qquad \begin{array}{r} \boxed{9} \\ + 8 \\ \hline 17 \end{array}$$

SRB 108

4 What time does the clock show?

7:15

5 **Writing/Reasoning** Explain how you found the numbers that were 10 more and 10 less in Problem 1.

I minus 10 and add 10

SRB 89, 98

How Long is Math Class?

Lesson 1-6

DATE TIME

Math Message

Sheena's math class began at 9:55 A.M. and ended at 11:10 A.M. She started to figure out how long the class lasted. She used a number line.

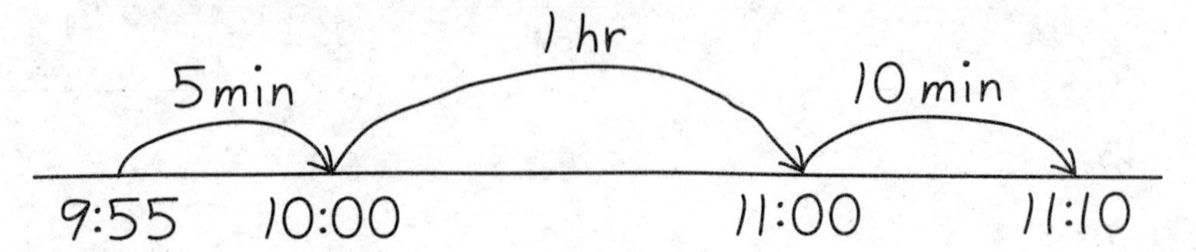

Use Sheena's number line to complete the problem. Tell the length of time in hours and minutes.

Math class lasted 1 hour and 15 minutes.

Explain Sheena's strategy to your partner.

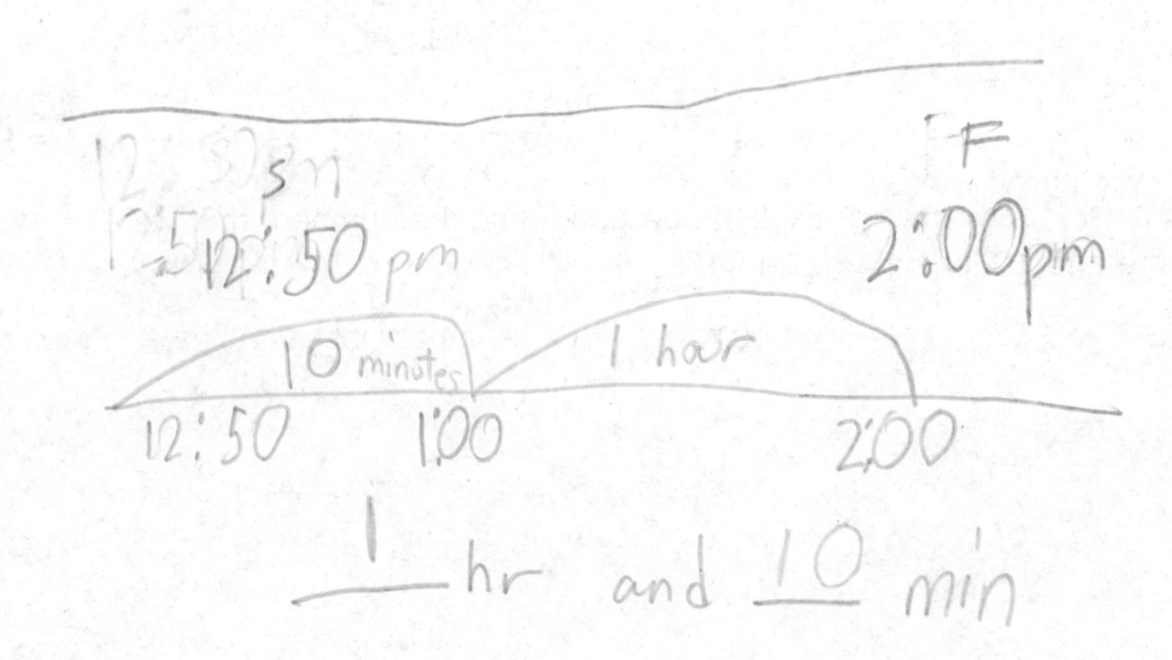

Math Boxes

1 Write the number that is halfway between 30 and 40 on the number line.

Then write 37 where it belongs on the number line.

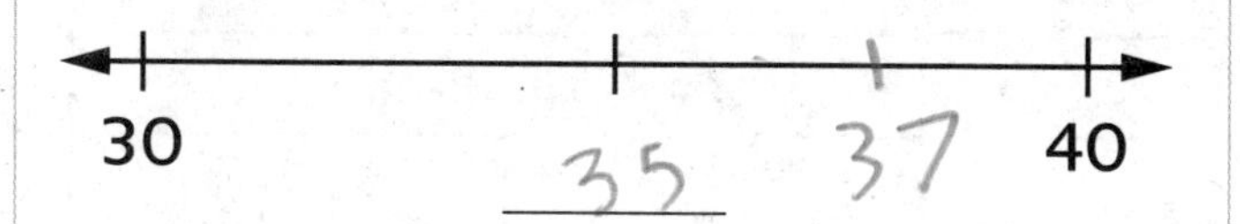

35 37

37 rounded to the nearest 10 is 40.

2 You may draw open number lines to help.

34 rounded to the nearest 10 is 30.

29 rounded to the nearest 10 is 30.

3 Use your Pattern-Block Template. Trace a polygon with 4 equal sides.

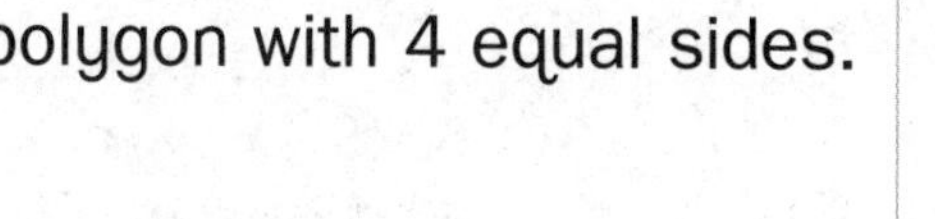

What shape did you draw?

square

4 Find the difference between each pair of numbers. You may use the number grid.

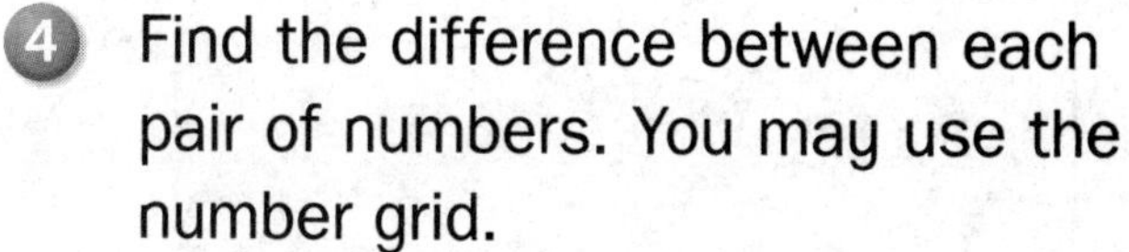

52 and 75 23

20 and 84 64

27 and 60 33

5 Draw the minute and hour hands to show 8:45.

6 Write the number that is 10 more.

59 69

160 170

120 130

92 102

901 911

Displaying Data

Lesson 1-7

DATE TIME

1. How many last names are there? ________________

2. Use the data you collected to make a tally chart for the last names in your class. Add rows as needed.

Last Names	
Number of Letters	Number of Children

3. Look at the data in your tally chart. Write at least three things you know from looking at the data.

____

____

____

Displaying Data (continued)

Lesson 1-7

DATE TIME

4 Make a bar graph for your set of data.

Title: ______________________________

Math Boxes

Lesson 1-7

DATE TIME

Math Boxes

1 Write the numbers that are 10 less and 10 more.

10 less		10 more
39	49	59
346	356	366
399	409	419
788	798	808

SRB 89, 98

2 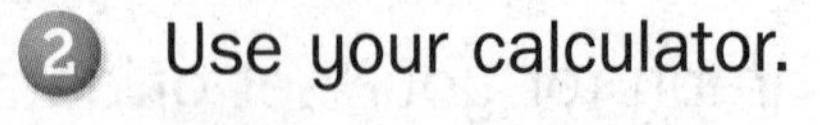

Use your calculator.

Enter	Change to	How?
29	49	+20
84	44	−40
188	208	+20
403	603	+200

SRB 293

3 Add.

Unit

16 = 8 + 8

6 + 7 = 13

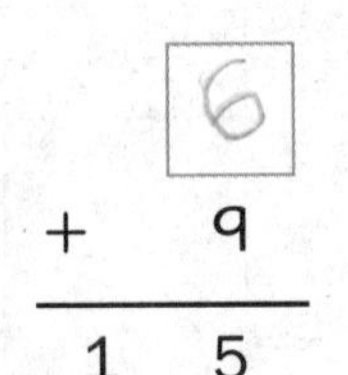

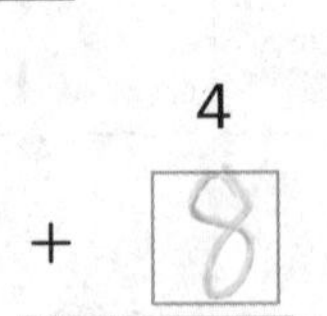

SRB 108

4 What time does the clock show?

11 : 40

SRB 186

5 **Writing/Reasoning** What time will it be 3 hours from the time in Problem 4? 2:40

Explain how you solved the problem.

I moved the hour 3 times and it said 2:40

SRB 18-19

Sharing Strategies for Equal Groups

Lesson 1-8

DATE TIME

Math Message

Solve. Include sketches to show your thinking.

Ellie bought 3 packs of stickers. There are 6 stickers in each pack. How many stickers did Ellie buy in all?

______ stickers

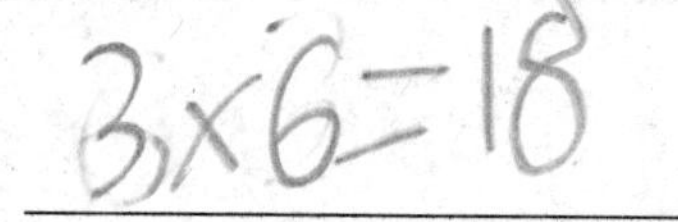

Number model: ______

1. Solve. Include sketches to show your thinking.

   Max keeps his baseballs in a rectangular box. The box fits 4 rows of baseballs with 5 balls in each row. How many baseballs can Max fit into his box?

______ baseballs

Number model: ______

2. For other number stories, draw sketches to show your solutions and write number models.

Story about:	Story about:
______	______
Number model:	Number model:
______	______

Writing Multiplication Number Stories

Lesson 1-8

DATE TIME

Choose a number sentence from the bank. Tell a number story to match your number sentence.

Number Sentence Bank		
$3 \times 4 = 12$	$2 \times 5 = 10$	$4 \times 2 = 8$
$5 \times 3 = 15$	$3 \times 6 = 18$	$4 \times 4 = 16$

My Number Sentence: $4 \times 4 = 16$

Write a number story to match your number sentence. Draw a picture of your story.

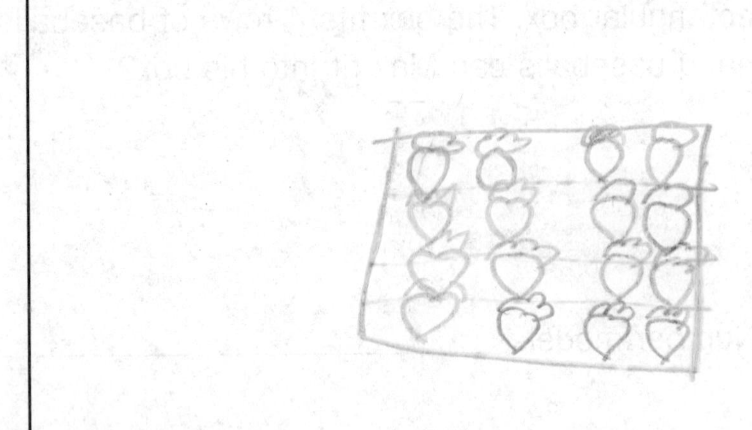

My mom got a box of strawberries.

DATE TIME

1. Write the number that is halfway between 50 and 60 on the number line.

Then write 52 where it belongs on the number line.

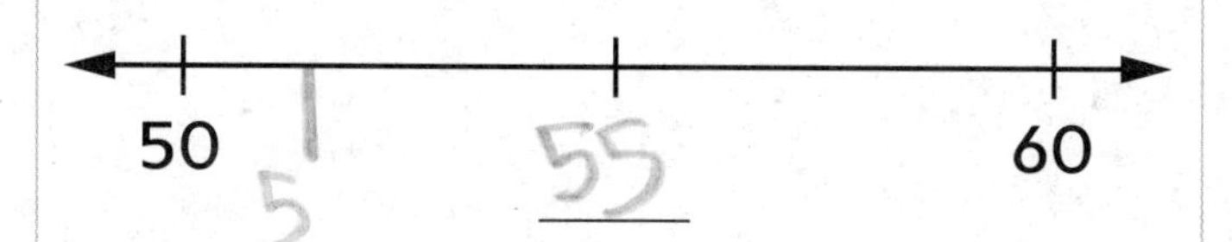

What is 52 rounded to the nearest 10?

50

SRB 104

2. You may draw open number lines to help.

286 rounded to the nearest 100 is 300.

593 rounded to the nearest 100 is 600.

SRB 104

3. Use your Pattern-Block Template. Trace a polygon with 4 sides that are not all the same length.

What shape did you draw?

parulalagram

SRB 105

4. Find the difference between each pair of numbers. You may use the number grid.

41 and 79 38

30 and 96 66

18 and 80 62

SRB 91

5. Record the time.

2:25

SRB 186

6. Write the numbers that are 10 less and 10 more.

SRB 89, 98

10 less		10 more
123	133	143
145	155	165
220	230	240
497	507	517
692	702	712

Equal Sharing and Equal Grouping

Lesson 1-9

DATE TIME

Math Message

Solve. Sketch to show your thinking.

Ms. Smith has 20 scissors. She places scissors on 5 tables. How many scissors can she place on each table?

$20 \div 5 = 4$

Answer: 4 scissors

1. Tomás is making snacks for his team. He has 15 strawberries and puts 3 strawberries in each teammate's bag. How many teammates does he have?

   Answer: 5 teammates

2. Listen to each story. Show your work with sketches.

Story about ______	Story about ______
Answer: ______ (unit)	Answer: ______ (unit)

Division Number Stories

Lesson 1-9

DATE TIME

Draw pictures to help you solve each problem below.
Record your answer.
Wait until your teacher tells you to write the number models.

1. Kate wants to equally share 12 snap cubes among 3 friends. How many cubes will each friend get?

Answer: 4 snap cubes

Number model: 12 ÷ 3 = 4

2. Ms. Early is making sets of cubes to share. She has 25 total cubes, and she puts them into sets of 5 cubes each. How many sets can she make?

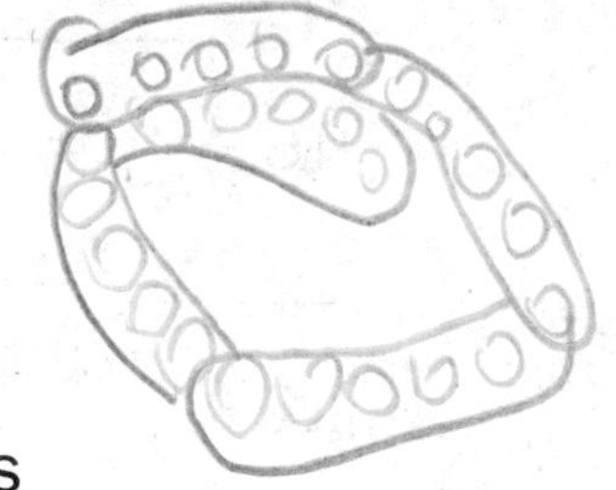

Answer: 5 sets

Number model: 25 ÷ 5 = 5

Try This

3. Manny wants to figure out the number of horses in a large stall. He can only see the horses' legs. He counts 28 legs. How many horses are there?

Answer: 7 horses

Number model: 28 ÷ 4 = 7

Math Boxes

Lesson 1-9

DATE TIME

1

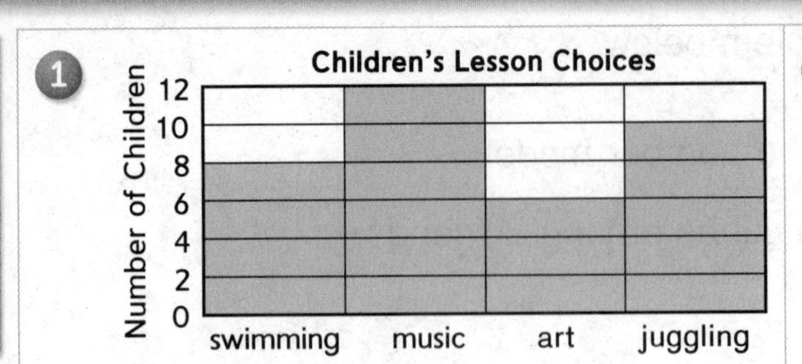

Did more children like music or swimming? music

How many more children chose music than art?

6

SRB 191

2 Cindy leaves her house at 9:00 A.M. She arrives at her grandmother's house at 11:00 A.M. How long does it take to get to her grandmother's house? Use your toolkit clock or draw an open number line to help.

2 hours
(units)

3 Estimate the length of this line segment to the nearest inch.

About 3 inches

Use a tool to check your estimated length.

SRB 171

4 Fill in the unit. Solve.

Unit

$9 = \_\_\_\_\_ - 6$

$16 - 8 = \_\_\_\_\_$

$\_\_\_\_\_ = 11 - 4$

$12 - \_\_\_\_\_ = 5$

$9 = \_\_\_\_\_ - 9$

5 **Writing/Reasoning** Explain what tool you used and how you used it to check your estimated length in Problem 3.

Dimes and Nickels Totals

1 Complete the table.

Number of dimes	Sketch	Multiplication number model
2 dimes	10¢ 10¢	$2 \times 10¢ = 20¢$
4 dimes	10¢ 10¢ 10¢ 10¢	4 x 10¢ = 40¢
5 dimes	10¢ 10¢ 10¢ 10¢ 10¢	5 x 10¢ = 50¢
10 dimes	10¢ 10¢ 10¢ 10¢ 10¢ 10¢ 10¢ 10¢ 10¢ 10¢	10 x 10¢ = $1.00

2 Complete the table.

Number of nickels	Sketch	Multiplication number model
2 nickels	5¢ 5¢	$2 \times 5¢ = 10¢$
4 nickels	5 5 5 5	4 x 5¢ = 20¢
5 nickels	5 5 5 5 5	5 x 5¢ = 25¢
10 nickels	5 5 5 5 5 5 5 5 5 5	10 x 5¢ = 50¢

Math Boxes

DATE TIME

Math Boxes

1 

There are 10 shoes in all.

Write a number model.

2×5=10

SRB 38

2 What time is it?

Fill in the circle next to the correct answer.

Ⓐ 6:45 Ⓑ 9:33

Ⓒ 10:35 Ⓓ 8:30

3 Add or subtract on your calculator to complete these problems.

Enter	Change to	How?
163	193	+30
603	803	+200
345	305	−40
341	41	−300

SRB 293

4 Write the numbers that are 100 less and 100 more.

100 less		100 more
313	413	513
402	502	602
632	732	832
791	891	991

5 Write the number that is halfway between 60 and 70 on the number line. Label 66 where it belongs.

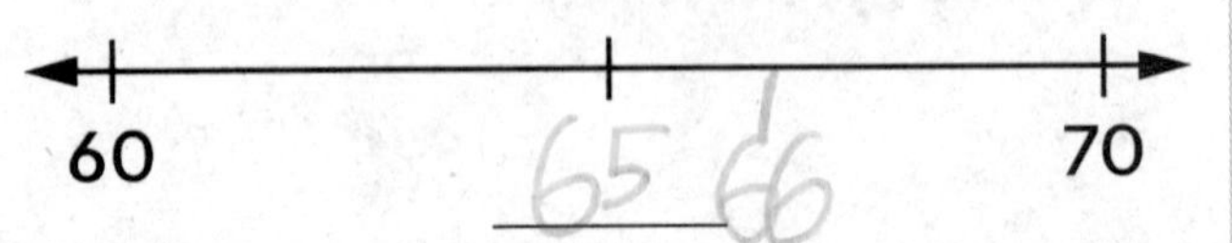

65 66

66 rounded to the nearest ten is 70.

6 Josh got to soccer practice at 3:10 P.M. He left at 3:55 P.M. How long was he at practice?

45 minutes (unit)

SRB 18-19

Sunrise and Sunset Data

Lesson 1-11
DATE TIME

Date	Time of Sunrise	Time of Sunset	Length of Day
			hr min
			hr min
			hr min
			hr min
			hr min
			hr min
			hr min
			hr min
			hr min
			hr min
			hr min
			hr min
			hr min
			hr min
			hr min
			hr min
			hr min
			hr min
			hr min
			hr min
			hr min
			hr min
			hr min
			hr min
			hr min

1. Fill in a unit. Solve.

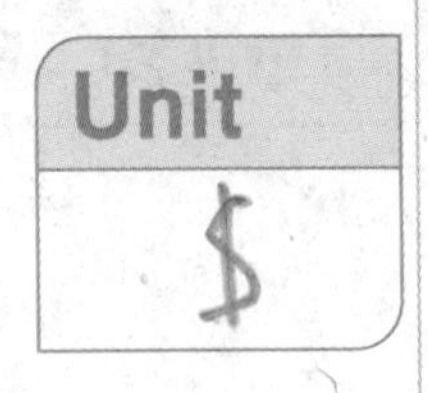

10 − 2 = 8

20 − 12 = 8

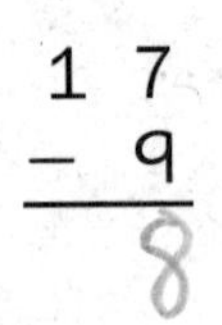

$$\begin{array}{r} 17 \\ -\ 9 \\ \hline 8 \end{array} \qquad \begin{array}{r} 27 \\ -19 \\ \hline 8 \end{array}$$

2. The library has 33 red pencils and 40 blue pencils. How many pencils in all? You may draw a diagram.

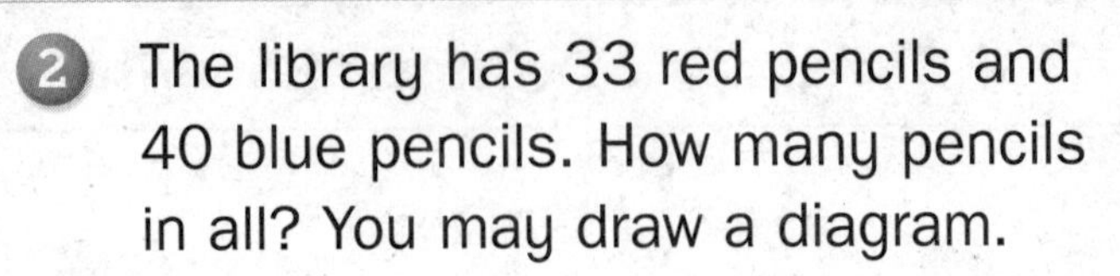

Answer: 73 pencils

Number model: 40+33=73

3.

How many dots in all? 20

Write a multiplication number sentence for the array.

4×5=20

SRB 41-42

4. Ty has 15 pennies to put into 3 groups. Draw a picture to show how many pennies are in each group.

There are 5 pennies in each group.

5. **Writing/Reasoning** For Problem 4, Ty wrote this number model: 15 ÷ 3 = 5. Do you agree with Ty? Explain.

Yes because you need to divide the pennies by 3 boxes so you need to divide 15 and 3.

Pan Balance and Mass Record

DATE TIME

Follow the directions on Activity Card 15.

Record your work below.

1
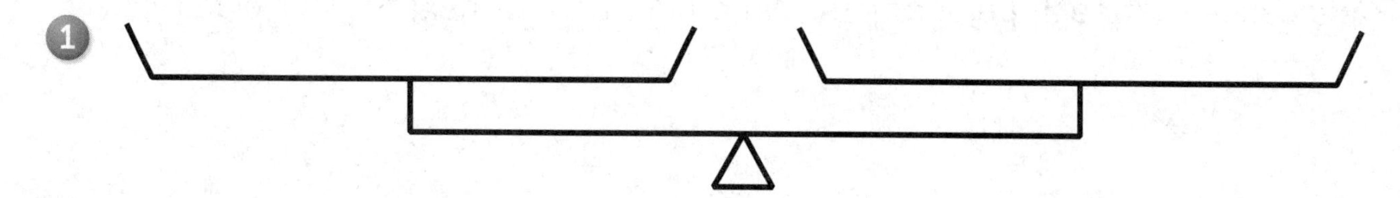

2
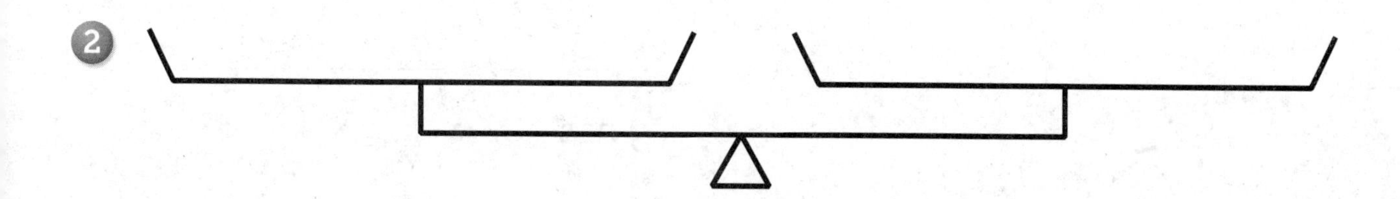

3
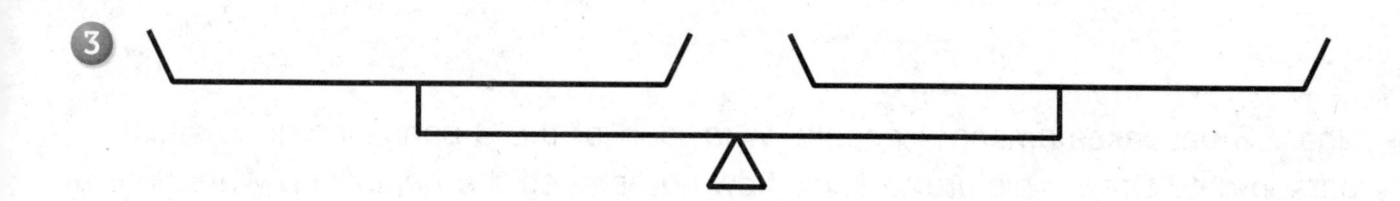

4
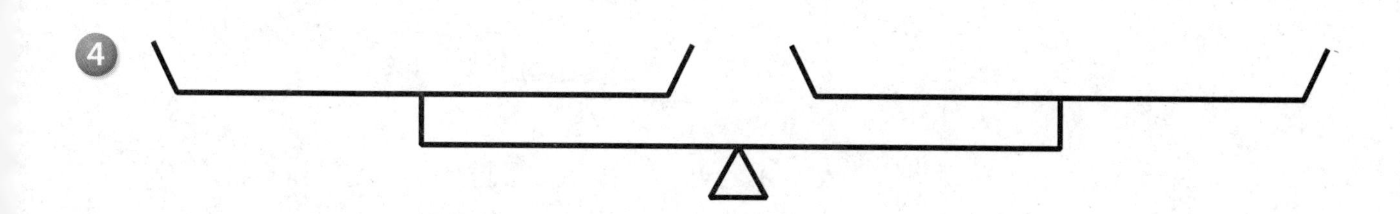

5
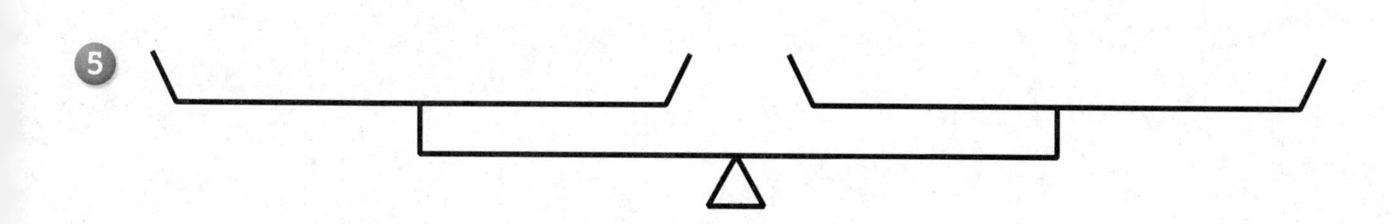

Equal Shares at a Pancake Breakfast

Lesson 1-12

DATE TIME

Follow the directions on Activity Card 16.

1. Share 3 pancakes equally among 6 people. Draw a picture to show part of the 3 pancakes that each person gets. Write your answer next to your picture.

2. Share 3 pancakes among 4 people. What part of the 3 pancakes does each person get? Draw a picture to show how you shared the pancakes. Write your answer next to your picture.

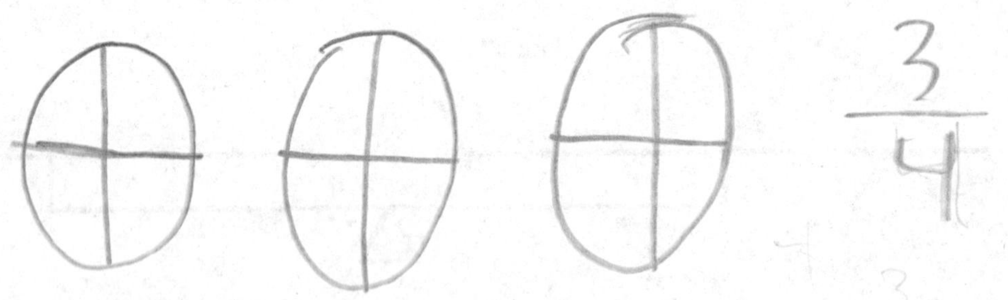

1

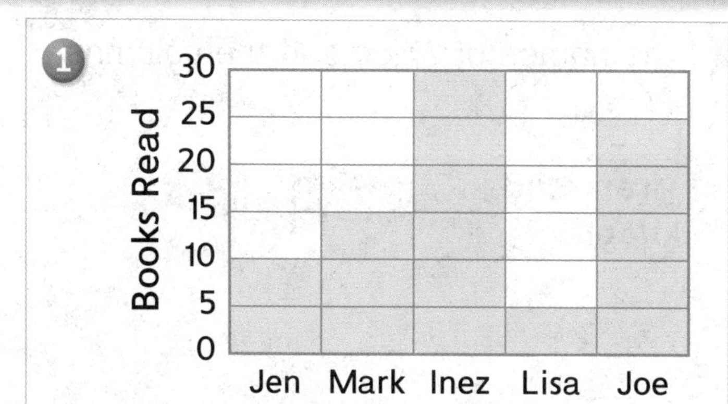

a. How many books did Jen and Mark read all together?

25 books

b. How many more books did Joe read than Lisa?

20 books

2 Jacob went to his friend's house at 8:30 A.M. He stayed there until 10:00 A.M. How long was he at his friend's house? Use your toolkit clock or draw an open number line.

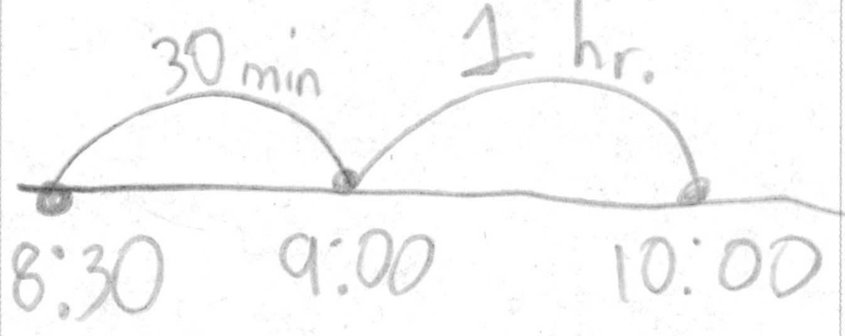

1 hr. 30 min

3 Find the difference between 91 and 59.

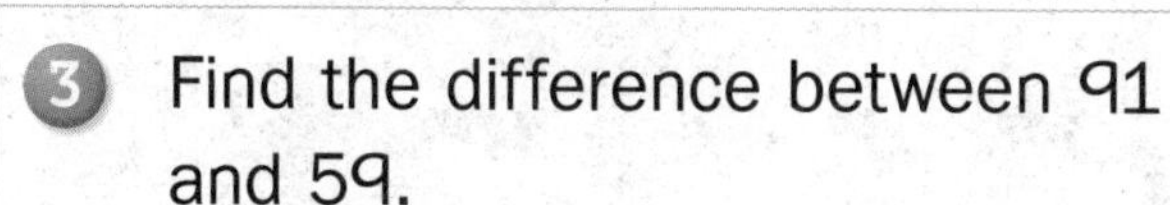

32

Which tool could help you check your answer?

Number Grid

4 Fill in the unit. Solve.

Unit

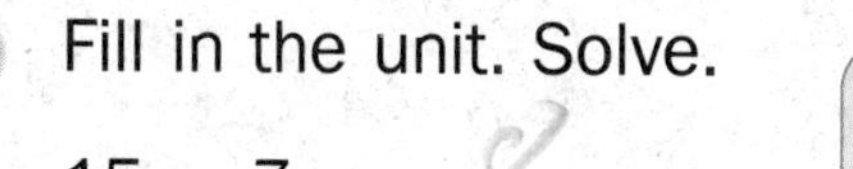

15 − 7 = 8

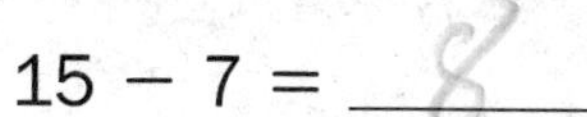

8 = 12 − 4

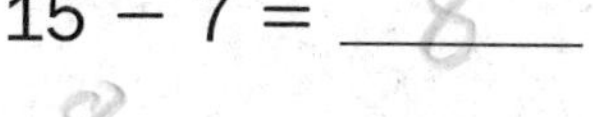

13 − 5 = 8

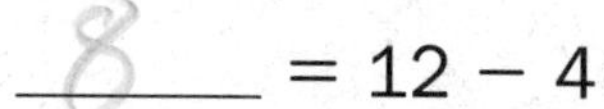

7 = 6 − 7

5 **Writing/Reasoning** Explain how you used the graph to answer Problem 1a.

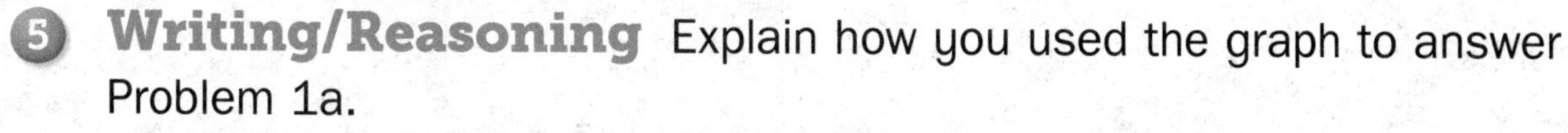

I picked Jen and Mark. And then

Measurement Hunt

Lesson 1-13

DATE TIME

1. Estimate the masses of objects. Record the names of objects in the columns below based on your estimates.

About 1 gram	More than 1 gram and less than 1 kilogram	About 1 kilogram

2. Use a pan balance and standard masses to find the actual masses of your objects. Record your work in the table below. Write the unit.

Name of Object	Mass

Mass Number Stories

Lesson 1-13

DATE TIME

Use mass measurements from *Student Reference Book,* page 271. Solve the number stories.

1. One soccer ball has a mass of about 425 grams. Dylan carries 2 soccer balls outside for gym class. What is the mass of 2 soccer balls together?

   425 x 2 = 850

   about 850 grams

2. One golf ball has a mass of about 43 grams. Keisha can juggle 3 golf balls. If she drops one golf ball, what is the mass of the remaining balls?

   43
   + 43
   86

   about 86 grams

3. Make up your own number story using the sports ball masses.

   One cricket ball is 159 grams. My friend can hold 3 balls. How many grams can my friend hold?

4. Trade papers with a partner and solve Problem 3. Show your work.

   159
   + 159
   + 159
   477 grams

Math Boxes

Lesson 1-13

DATE TIME

Math Boxes

1

There are 20 flower petals.

Number model: 4 x 5 = 20

SRB 38

2 Set your toolkit clock to 4:00. Then set it to 4:04 and draw the hands on the clock below.

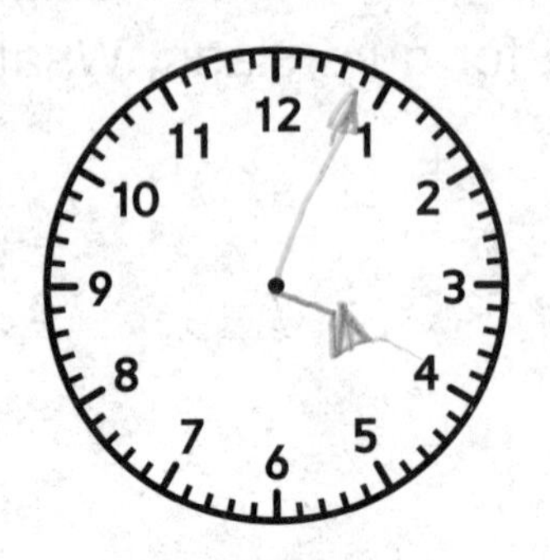

3 Add or subtract on your calculator to complete these problems.

Enter	Change to	How?
231	531	+300
756	696	-140
875	775	-100
985	485	-500

SRB 293

4 Write the numbers that are 100 less and 100 more.

100 less		100 more
108	208	308
299	399	499
554	654	754
707	807	907

5 Write the number that is halfway between 80 and 90 on the number line. Then write 83 where it belongs.

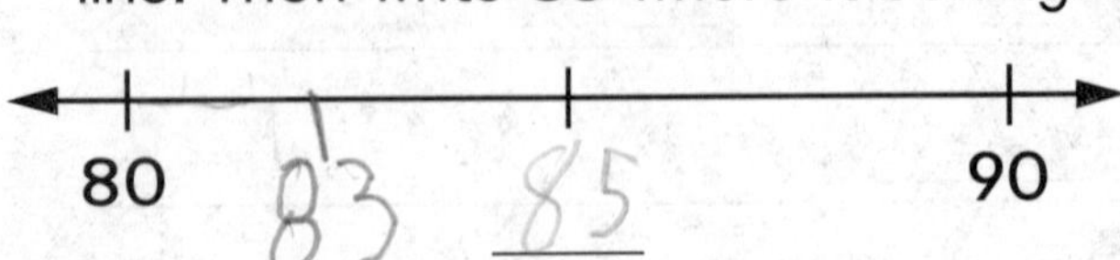

83 rounded to the nearest ten is 80.

SRB 104

6 The bus left at 8:30 A.M. It arrived at 9:30 A.M. How long was the ride?

Circle the best answer.

A. 1 hour

B. 50 minutes

C. 1 hour 10 minutes

D. 40 minutes

Math Boxes
Preview for Unit 2

1 Fill in a unit. Solve.

19 = 13 + 6

29 = 23 + 6

20 = 13 + 7

40 = 33 + 7

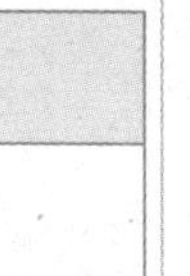

2 Claire had $40 in her bank. She spent $16. How much does she have left? You may draw a diagram.

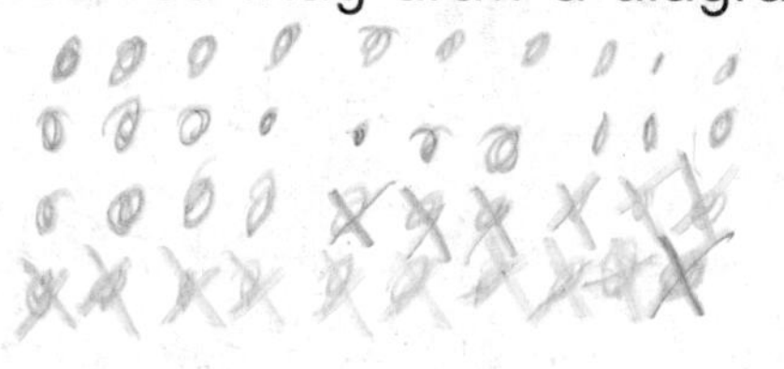

Answer: 34 (unit)

Number model:

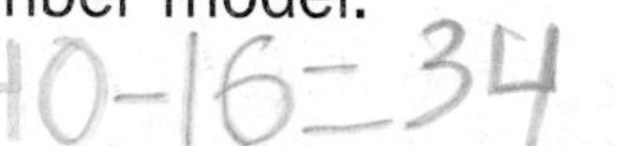

3

How many dots in all? 25

Write a multiplication number sentence.

5x5=25

4 4 children share 24 pennies equally. How many pennies does each child get? Use counters or draw a picture to help.

Answer: 6 (unit)

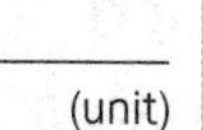

5 **Writing/Reasoning** Nicholas wrote 5 + 5 = 10 as a number sentence for Problem 3. Do you agree? Explain your answer.

No because, 5+5=10 but, the answer is 25 so, Nicholas can do 5+5+5+5+5=25 or 5x5=25.

Using Basic Facts to Solve Fact Extensions

Lesson 2-1

DATE TIME

Fill in the unit box. Complete the fact extensions.

Unit

1

5 = 12 − 7

50 = 120 − 70

500 = 1,200 − 700

2

8 + 3 = 11

80 + 30 = 110

800 + 300 = 1,100

Complete the fact extensions.

3

14 = 6 + 8

24 = 16 + 8

64 = 56 + 8

4

14 − 9 = 5

24 − 9 = 15

54 − 9 = 45

5 Explain how you used a basic fact to help you solve Problem 4.

If I know 14−9=5,

Add or subtract to complete these problems on your calculator. You may also use a number grid or base-10 blocks.

6

Enter	Change to	How?
33	40	+7
80	73	−7
80	23	−57

7

Enter	Change to	How?
430	500	+70
700	640	−60
1,000	400	−600

8 What combination of 10 helped you solve Problem 6? Explain.

3+7=10

Finding Elapsed Time

Lesson 2-1

DATE TIME

Use a toolkit clock or an open number line to help you solve these problems. Show your work.

1. Ava leaves to go swimming at 4:05 and returns at 4:57. How long has she been gone?

50 min 2 min

4:05 4:55 4:57 52 min

2. Deven rides his bike 4 miles. He rides from 10:15 A.M. until 11:20 A.M. How long does it take him to ride 4 miles?

5 min

Try This

3. LaToya leaves for school at the time shown on the first clock. She returns home at the time shown on the second clock. How long is LaToya away from home?

7 hr. 55 min.

Explain how you figured out the length of LaToya's school day.

I put my clock on 8:20. I counted when I move the clock to 4:15 I thinked and I said to myself, "It is 7hrs and 55mins.

Math Boxes

Lesson 2-1

DATE TIME

Math Boxes

1 Karen arrived at her cousin's house at 12:30 P.M. and left at 6:30 P.M. How long did she stay? You may use a clock or an open number line to help.

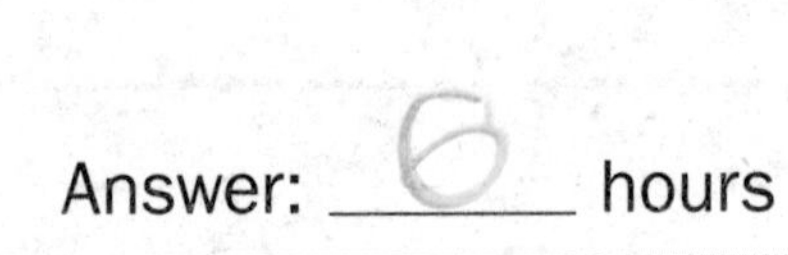

Answer: ______ hours

SRB 18-19, 187-188

2 Complete the Fact Triangle. Write the fact family.

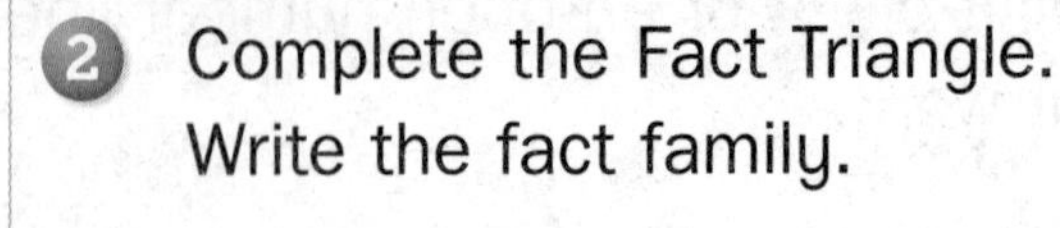

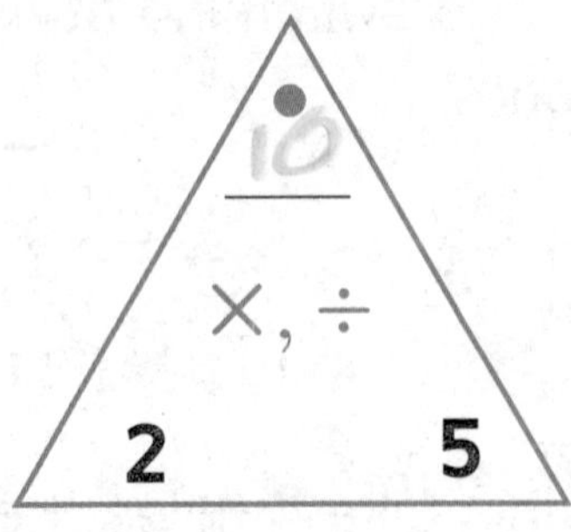

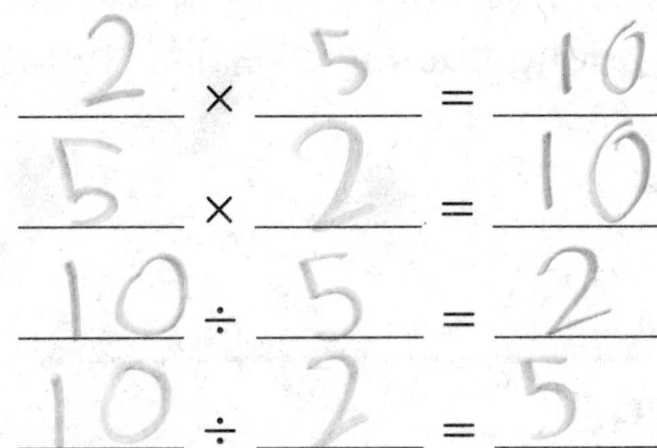

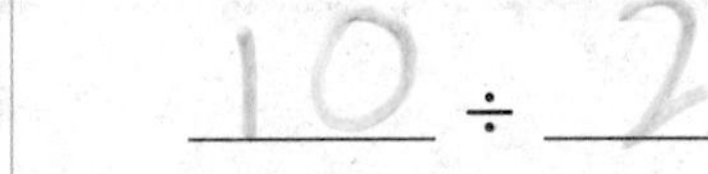

3 Record the time.

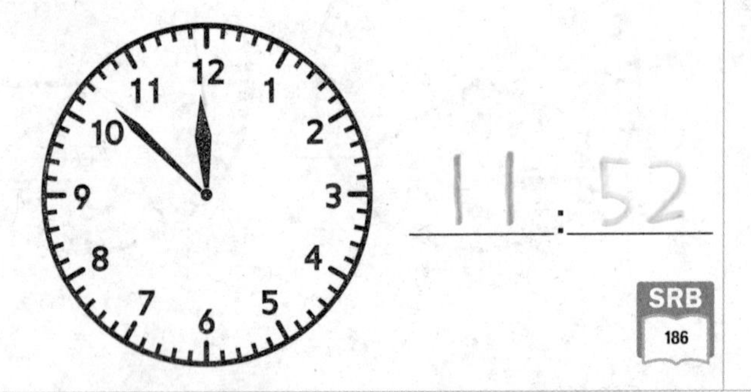

____ : ____

SRB 186

4

There are ______ legs in all.

Write a multiplication number model: ______________

5 **Writing/Reasoning** Jamal completed the Fact Triangle in Problem 2 as shown. Is he correct? Explain using words or a picture.

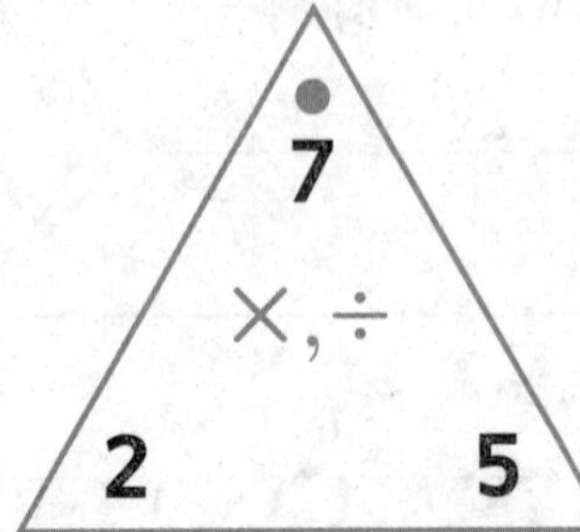

Number Stories

DATE TIME

For each number story:

- Write a number model. Use ? for the unknown.
- You may draw a diagram like those shown below or a picture to help.

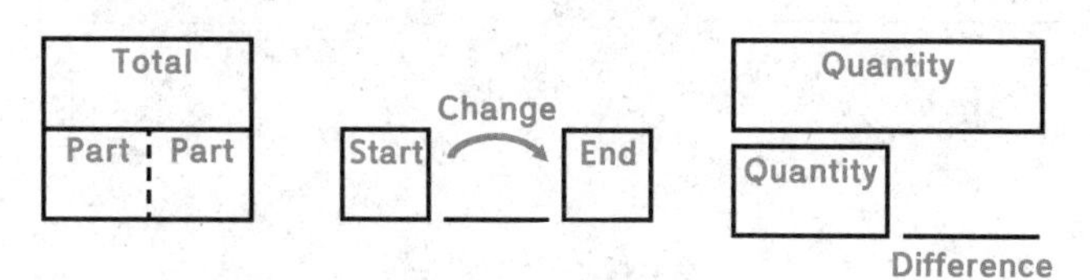

- Solve the number story and write your answer.
- Explain how you know your answer makes sense.

1 Two pythons each laid a clutch of eggs. One clutch had 47 eggs. The other had 32 eggs. How many eggs were there in all?

Number model: 47+32=?

Answer the question: 79 eggs (unit)

Check: How do you know your answer makes sense?

I added the ones 7+2=9, then I added the tens 40+30=70. 70+9=79

2 An alligator clutch had 60 eggs. Only 12 hatched. How many eggs did not hatch?

Number model: 60-12=?

Answer the question: 48 eggs (unit)

Check: How do you know your answer makes sense?

I borowed 1 ten from 60, so in the ones place, it should 10 and in the tens place.

Number Stories (continued)

3 Ahmed had $22 in his bank account. For his birthday, his grandmother deposited $25 for him. How much money is in his bank account now?

Unit
dollars

Number model: 22+25=?

Answer the question: $47

Check: How do you know your answer makes sense?

I added 2+5 and that equal 7 and 2+2 is 4 so, 40+7=47.

4 Omar had $53 in his piggy bank. He used $16 to take his sister to the movies and buy treats. How much money is left in his piggy bank?

Number model: 53-16=?

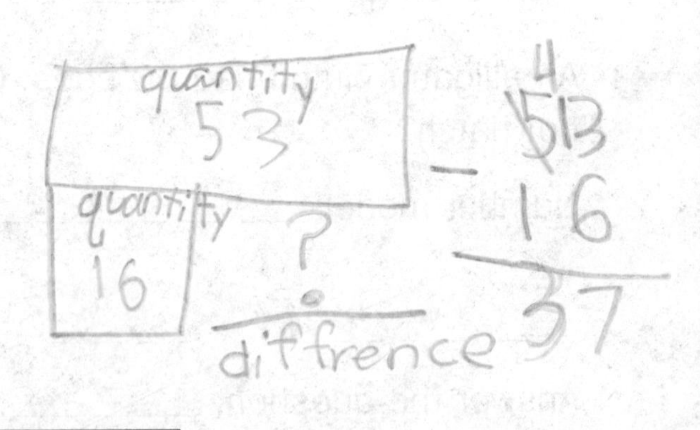

Answer the question: $37

Check: How do you know your answer makes sense?

I borrowed 5 and in the ones Place, it is 13 and in the tens place, it is 4 so, 13-6=7 and 4-1=3 so 30+7=37.

Math Boxes

Lesson 2-2

DATE TIME

1 Choose the best answer.

The mass of a centimeter cube is about

(A) 1 gram.

(B) 10 grams.

(C) 50 grams.

(D) 100 grams.

2 Round each number to the nearest 10. You may draw open number lines to help.

68 70

83 80

3 Rachel has 3 packages of seeds. Each package has 10 seeds. How many seeds does she have in all?

Answer: 30 seeds

Write a multiplication number model. $3 \times 10 = 30$

4

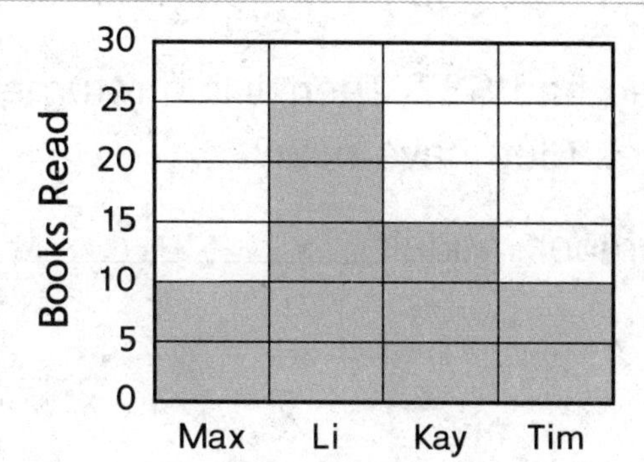

How many more books did Li read than Max? 15 books

How many books did Kay and Tim read all together? 25 books

5 **Writing/Reasoning** Write a number story to fit $2 \times 4 = 8$.

We have 2 pakages of books. Each package has 4 books. How many books are they

Explain how your number story fits $2 \times 4 = 8$.

More Number Stories

Lesson 2-3

DATE TIME

For each number story:

- Write a number model. Use a ? for the unknown.
- You may draw a diagram like the ones below or a picture to help you solve.

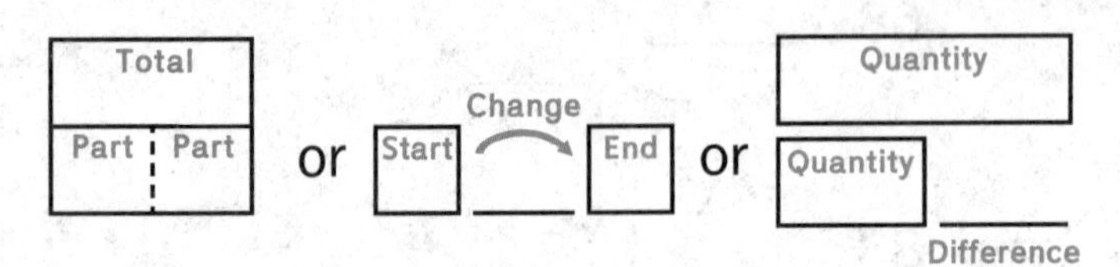

- Solve and write your answer with the unit.
- Explain how you know your answer makes sense.

1. Cleo had \$37. Then Jillian returned \$9 that she borrowed. How much money does Cleo have now?

   Number model: 37 + 9 = ?

   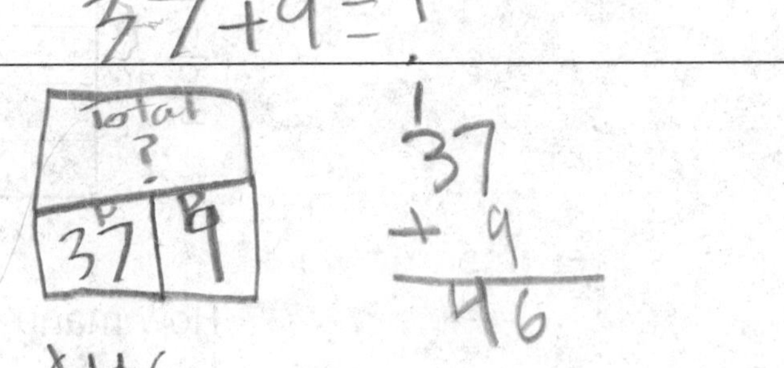

   Answer the question: 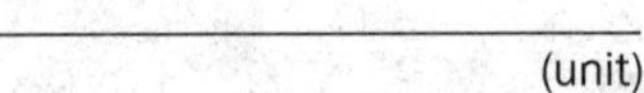 \$46 (unit)

   Check: How do you know your answer makes sense?

   ____

2. Audrey had \$61 in her bank account. She took out \$48. How much is left in her account?

   Number model: 61 - 48 = ?

   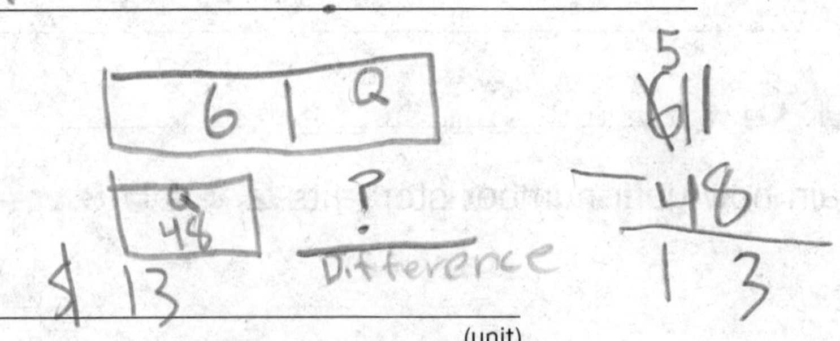

   Answer the question: \$13 (unit)

   Check: How do you know your answer makes sense?

   ____

More Number Stories (continued)

3. Pedro had 70¢. He bought grape juice and had 25¢ left. How much did the grape juice cost?

Number model: 70 − 25 = ?

  6
  ~~7~~10
− 25
  45

Answer the question: 45¢
(unit)

Check: How do you know your answer makes sense?

4. Nikhil had $40 in his wallet when he went to the carnival. When he got home, he had $18. How much did he spend at the carnival?

Number model: 40 − 18 = ?

  3
  ~~4~~10
− 18
  22

Answer the question: $22
(unit)

Check: How do you know your answer makes sense?

Math Boxes

Lesson 2-3

DATE TIME

Math Boxes

1 Chip started a bike ride at 7:00 A.M. and finished at 11:30 A.M. How long did he ride his bike? You may draw a clock or an open number line to help.

4 hrs 30 min
7:00 11:00 11:30

Answer: 4 hours 30 minutes

SRB 18-19, 187-188

2 Complete the Fact Triangle. Write the fact family.

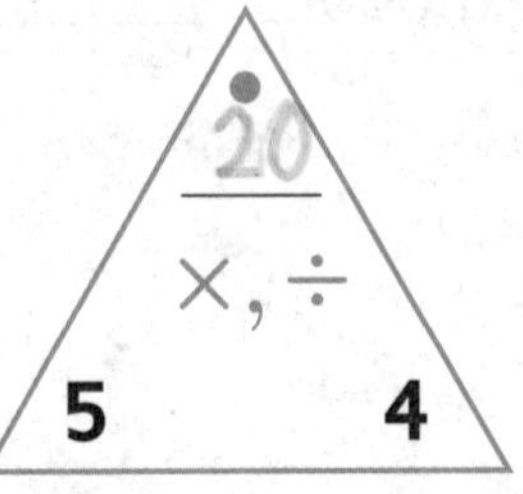

5 = 4 × 20

4 = 5 × 20

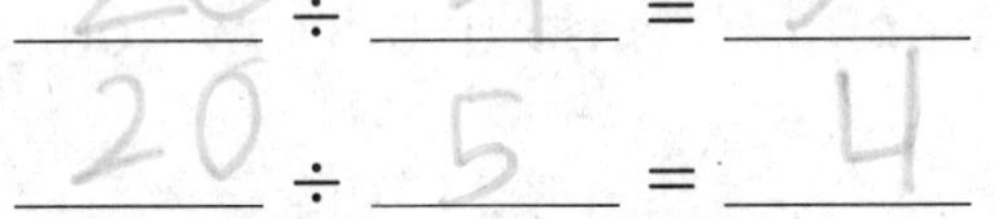

20 ÷ 4 = 5

20 ÷ 5 = 4

3 Draw the hands to show 6:23. You may set your toolkit clock to a familiar time to help.

4

There are 16 crayons in all.

Write a multiplication number model:

2×8=16

5 **Writing/Reasoning** Marla says that the number model 8 + 8 = 16 also matches the number story in Problem 4. Is she correct? Explain.

She is correct because they have 2 8s so she is correct

Multistep Number Stories, Part 1

Solve each problem below. Use pictures, words, or numbers to keep track of your thinking. Write number models to show each of your steps.

A package of rice cakes contains 6 rice cakes.

1. You buy 2 packages of rice cakes and then eat 4 rice cakes.
   How many rice cakes are left?

   Number models: $2 \times 6 = 12$ $\quad 12 - 4 = 8$

   Answer: 8 rice cakes

2. You buy 5 packages of rice cakes.
   You give 15 rice cakes away.
   How many rice cakes do you have now?

   Number models: $5 \times 6 = 30$ $\quad 30 - 15 = 15$

   Answer: 15 rice cakes

Try This

3. For everyone in your class to have one rice cake, how many packages would you need?

   21 children 4 packages

   Would there be any leftover rice cakes? 3

   How many? 24 rice cakes

   Number models: $4 \times 6 = 24$ $\quad 24 - 21 = 3$

Rounding Numbers

DATE TIME

Round the following numbers. Show your work on the open number lines.

1. What is 64 rounded to the **nearest 10**? 60

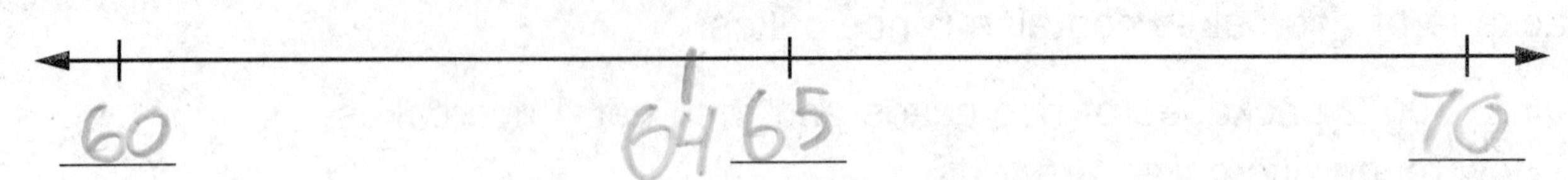

2. What is 278 rounded to the **nearest 100**? 300

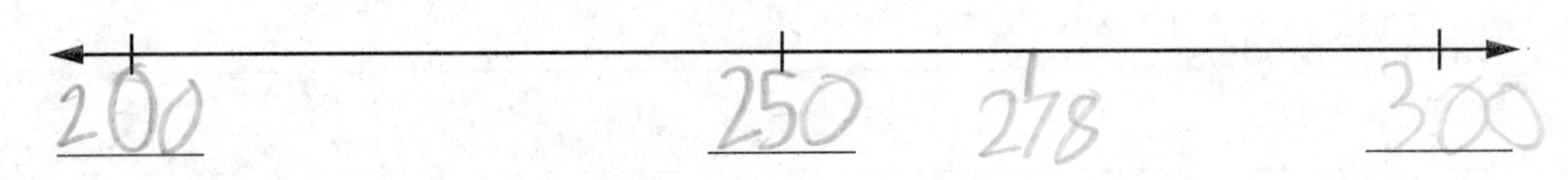

Here is another way to think of rounding numbers.

Round 27 to the nearest 10.

What multiples of 10 are close to 27? 20 and 30

What number is halfway between 20 and 30? 25

The halfway number is written at the top of the hill.

Would 27 be heading toward 20 or 30? 30

27 rounded to the **nearest 10** is 30.

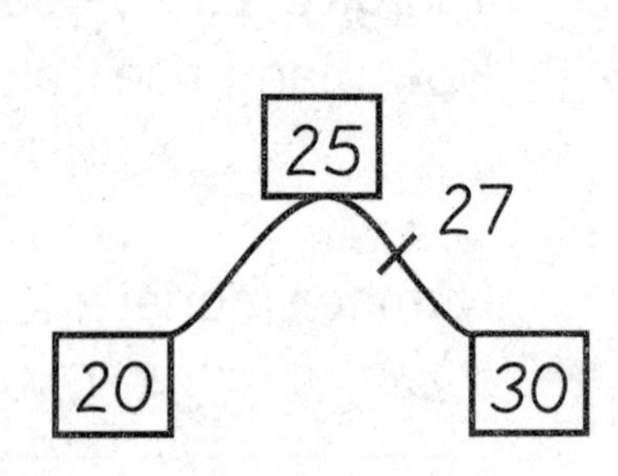

3. Fill in the "hill" to show how you would round 82 to the nearest 10.

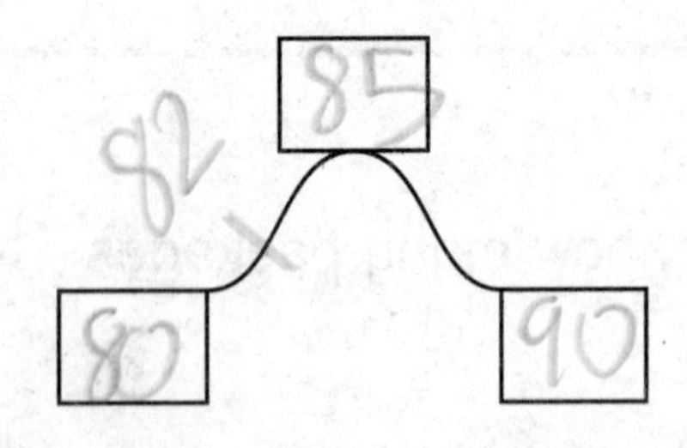

82 rounded to the **nearest 10** is 80.

4. Use either an open number line or hill to round 140 to the nearest 100.

140 rounded to the nearest 100 is 100.

Explain your work. In the tens

150
100
200

Math Boxes

1 Circle the standard mass you would use to measure the mass of each item.

centimeter cube

(1 gram) 1 kilogram

a 1-liter bottle of water

1 gram (1 kilogram)

2 Round each number to the nearest 100. You may draw open number lines to help.

224 ___200___

576 ___600___

3 Jalen has 5 bags with 10 apples in each bag. How many apples does he have in all?

Answer: ___50 apples___ (unit)

Fill in the circles next to the number models that fit the story.

(A) 5 + 10 = 15

(B) 5 × 10 = 50

(C) 10 + 10 + 10 + 10 + 10 = 50

(D) 5 + 5 + 5 + 5 + 5 = 25

SRB 38

4

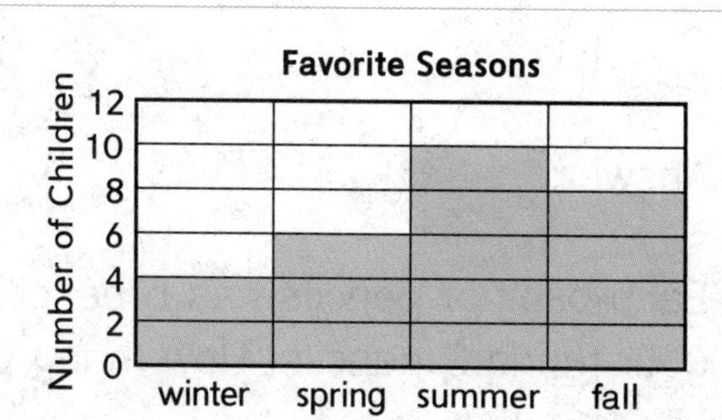

How many children in all chose summer and fall?

___18___ children

How many more children chose summer than spring?

___4___ children

5 **Writing/Reasoning** Write a number story to fit this number sentence: 7 × 2 = 14.

I have 7 bags of apples. In each bag they have 2 apples. How many apples are they in all?

Explain how your number story fits 7 × 2 = 14.

I know that because they have 7 groups of 2.

SRB 41-43

More Number Stories

Lesson 2-5

DATE TIME

Solve each problem below. Show your work with pictures, numbers, or words. Write number models to keep track of your thinking.

1. Jill has 83¢. She buys 2 erasers for 25¢ each. How much money does she have left?

   Number models: ____________________

   ____

   Answer: __________ ¢

2. Each pack of pencils has 5 pencils. You have 4 packs of pencils. Then you give your friend 2 pencils. How many pencils do you have left?

   Number models: ____________________

   ____

   Answer: __________ pencils

3. Three friends equally share 15 almonds. One friend eats 3 of her almonds. How many almonds does she have left?

   Number models: ____________________

   ____

   Answer: __________ almonds

Math Boxes

Lesson 2-5

DATE TIME

Math Boxes

1 What basic fact could help you solve 1,000 − 800?

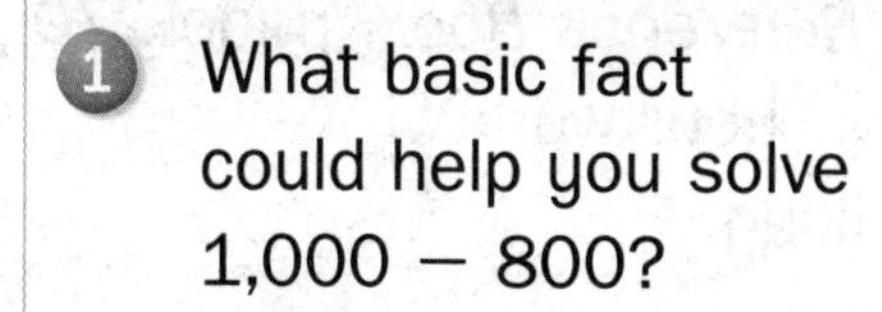

Unit
pens

Basic fact: 10 − 8 = 2

1,000 − 800 = 200

2 Solve. Corey had $75. He spent some money and now has $45. How much did he spend? You may draw a diagram or a picture.

75 − 45 = ?

(number model with ?)

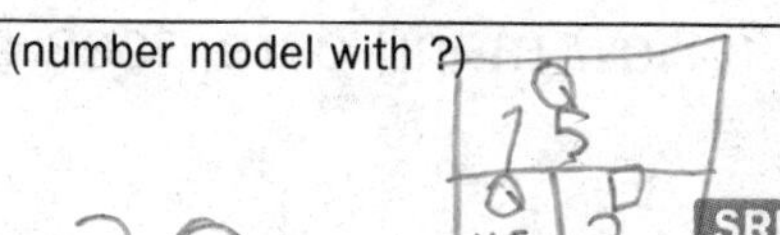

Answer: $ 30

3 Bradley shares 20 pencils among 5 tables. Draw a picture to show how many pencils are on each table.

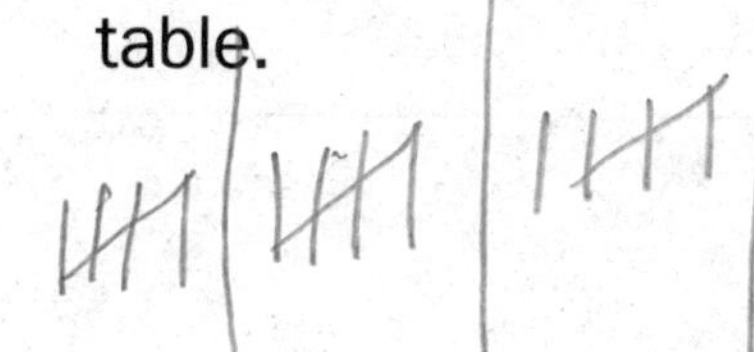

Answer: 4 pencils

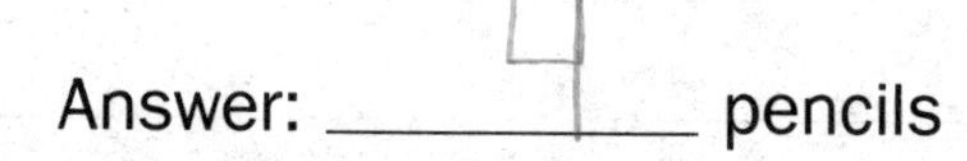

4 Divide the circle into two equal parts.

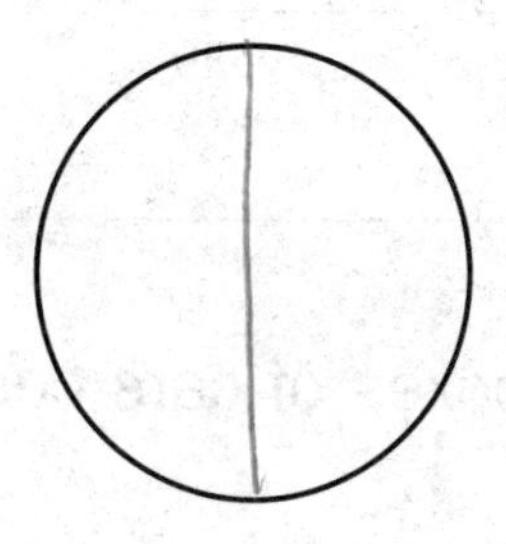

Use words to name 1 part.

one-half

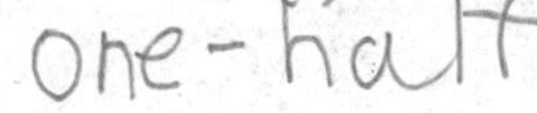

Use words to name all the parts together.

whole

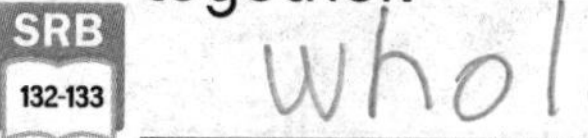

5 Andrea left her house at 1:15 P.M. for a walk. She returned at 1:50 P.M.

How long was her walk? 35 min Use the clock or the open number line to solve.

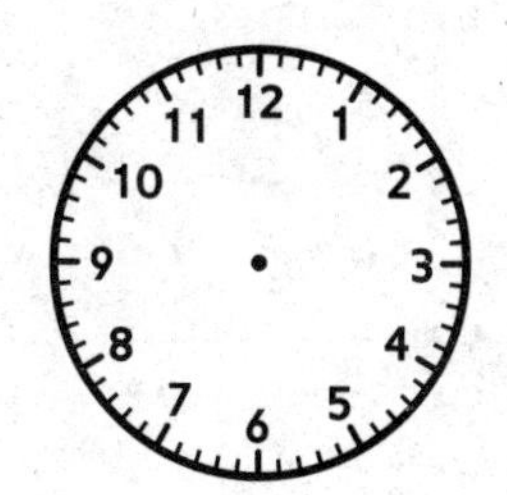

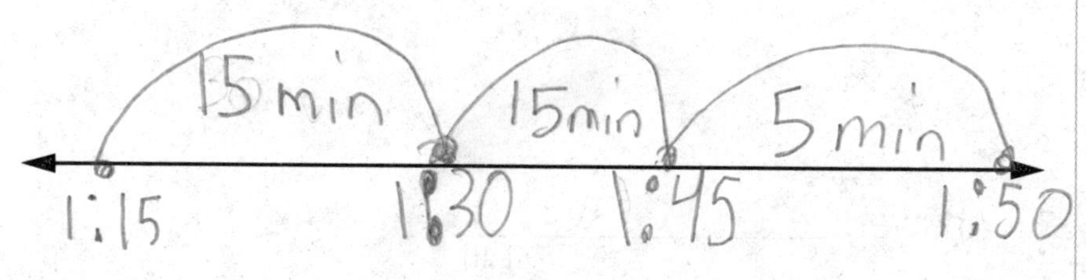

Equal-Groups Number Stories

Lesson 2-6

DATE TIME

Use the information on the Dollar Store Poster on *Student Reference Book*, page 270 to solve each number story. Use an efficient strategy. Show your work with drawings or words. Write number models to show your thinking.

1. Shanna buys 3 boxes of mini stock cars to share with her classmates. How many cars does she have all together?

   Answer: ______________ (unit)

   Number model: ______________________________

   How much do 3 boxes of cars cost?

   Answer: ______________ (unit)

   Number model: ______________________________

2. A teacher buys 1 package of Value Pack pens and 0 packages of chocolate-scented pens. How many pens does she buy in all?

   Answer: ______________ (unit)

   Number models: ______________________________

   ____

Math Boxes

DATE TIME

1 Austin read a book for 45 minutes on Monday and 25 minutes on Tuesday. How many more minutes did he read on Monday? You may draw a diagram or picture to help.

45-25=?

(number model with ?)

Answer: 20 minutes

2 Molly has two tennis balls. One ball has a mass of about 57 grams. What is the mass of two tennis balls?

$$\begin{array}{r} 57 \\ +\ 57 \\ \hline 114 \end{array}$$

About 114 grams

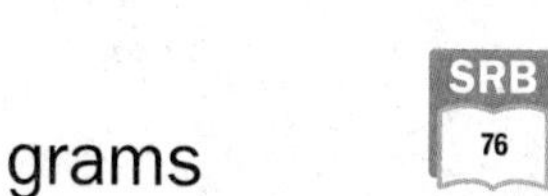

3 While playing *Multiplication Draw,* you roll a 5 and draw a 7 card.

Write a multiplication number sentence to record your turn.

5×7=35

4 Solve.

$4 \times 2 =$ 8

$2 \times 7 =$ 14

$5 \times 3 =$ 15

40 $= 4 \times 10$

20 $= 10 \times 2$

$6 \times 5 =$ 30

5 **Writing/Reasoning** On your next *Multiplication Draw* turn, you roll a 5 and draw an 8 card. Is your score, or product, greater or less than your score in Problem 3? Explain.

It is greater because problem 3's answer is 35 and problem5's answer is 40 so 40is greater than 35.

Representing Number Stories with Arrays

Lesson 2-7

DATE TIME

For each story, write the topic. Then write a number model with a ? for the unknown and draw an array to help solve.

1. Topic: crayons

   Number model: $3 \times 5 = 15$

   Answer: 15 crayons

2. Topic: fishs

   Number model: $6 \times 3 = 18$

   Answer: 18 fish

3. Topic: books

   Number model: $5 \times 8 = 40$

   Answer: 40 books

4. Topic: fruits

   Number model: $3 \times 2 = 6$

   Answer: 6 fruits

Math Boxes

Math Boxes

1 Fill in the unit. Solve.

Unit

17 − 8 = 9

170 − 80 = 90

1,700 − 800 = 900

SRB 114

2 Tiara played violin for 50 minutes in all. She played 35 minutes on Wednesday and played some more on Thursday. How many minutes did she play on Thursday? You may draw a diagram or picture to help.

50−35=?

(number model with ?)

Answer: 15 minutes

3 Dakota divides an 18-inch leather strip into 3 equal pieces to make bracelets. How long is each piece? You may draw a picture.

Fill in the circle next to the correct answer.

Ⓐ 4 inches

Ⓑ 6 inches

Ⓒ 8 inches

Ⓓ 10 inches

4 Divide the rectangle into fourths (4 equal parts).

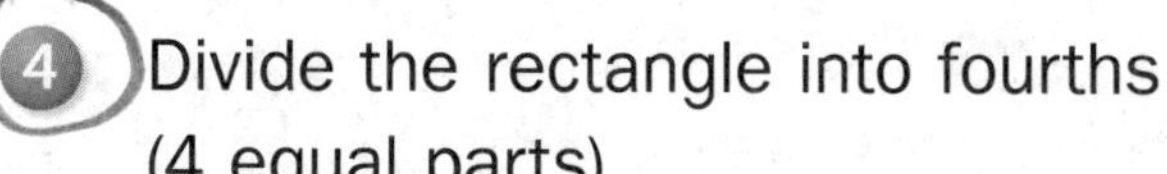

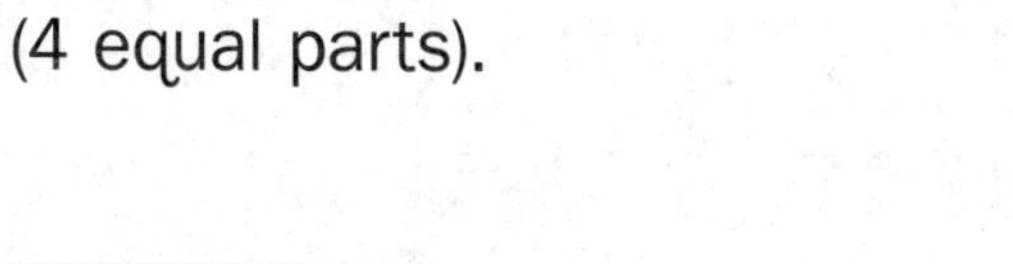

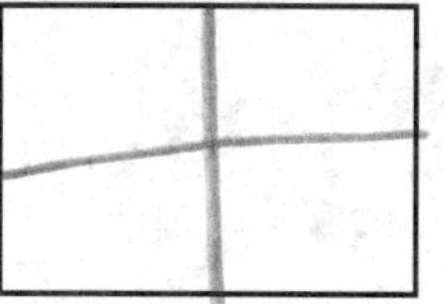

Use words(s) to name 1 part.

share

Use words(s) to name all the parts together.

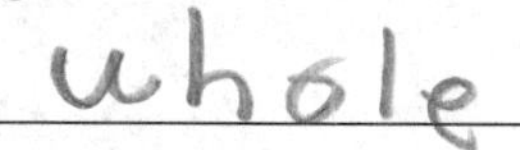

whole

5 Jamie leaves her house at 8:05 A.M. and arrives at her friend's house at 8:50 A.M. How long does it take her to get to her friend's house?

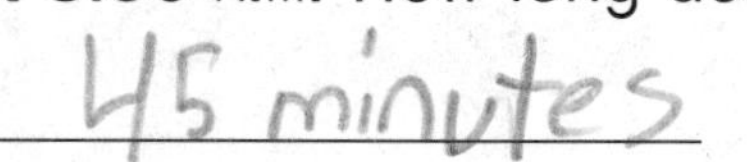

45 minutes

Use the clock or open number line to solve.

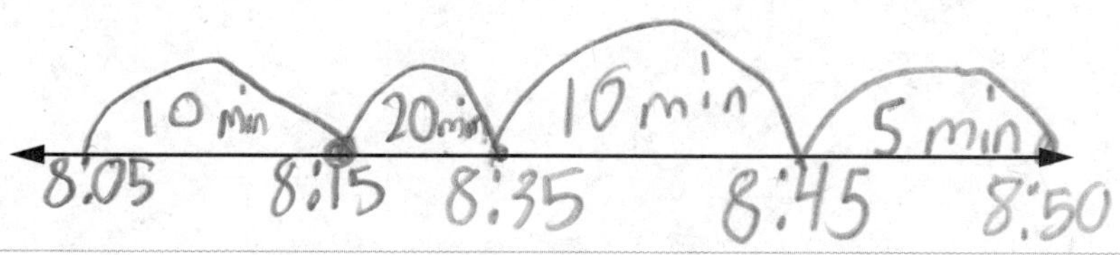

Sharing Pennies

Lesson 2-8

DATE TIME

Solve the problem below.
Use drawings, numbers, and words to show your thinking.

Leah and Matthew share 14 pennies equally.
How many pennies does each child get?

IIIIIII | IIIIIII

7 | 7

14 ÷ 2 = 7

7 × 2 = 14

14 − 7 = 7

7 + 7 = 14

I made 14 lines and I equaled 7 lines and the number sentences are 14 ÷ 2 = 7, 7 × 2 = 14, 14 − 7 = 7, and 7 + 7 = 14.

2 math vocabulary words

Answer: 7 pennies

Math Boxes

DATE TIME

1 Scientists counted 91 eggs in 2 clutches of python eggs. If 1 python clutch has 52 eggs, how many are in the other clutch? You may draw a diagram or picture.

91 − 52 = ?

(number model with ?)

Answer: 39 eggs

2 One golf ball has a mass of about 43 grams.

What is the mass of 3 golf balls together?

About 129 grams

3 James rolls a 2 and draws a 9 card in *Multiplication Draw.* Lucy rolls a 5 and draws a 3 card. Who has the larger product?

James

Write multiplication number sentences to record their turns.

2 × 9 = 18

5 × 3 = 15

SRB 248

4 Solve.

$1 \times 2 =$ 2

45 $= 9 \times 5$

$3 \times 2 =$ 6

$10 \times 3 =$ 30

25 $= 5 \times 5$

$7 \times 10 =$ 70

SRB 44

5 **Writing/Reasoning** Explain how you solved Problem 1.

I had 91 eggs. I borrowed 9 and in the ones place, it is 11 and in the tens place, it is 8. I know that 11-2=9 and 8-3=5 so, 30+9= 39.

Equal-Sharing Number Stories

Lesson 2-9

DATE TIME

Solve each number story and draw a sketch to show your thinking. Write a division number model for each story.

1. A class of 30 children wants to play ball. How many teams can be made with exactly 6 children on each team?

Answer: 5 teams

How many children are left over? 0 children

Number model: 30 ÷ 6 = 5

2. For another game, the same class of 30 children wants to have exactly 4 children on each team. How many teams can be made?

Answer: 7 teams

How many children are left over? 2 children

Number model: 30 ÷ 4 = 7 R2

3. Roberto has 25 pencils to share equally among 3 pencil boxes. How many pencils does he put in each box?

Answer: 8 pencils

How many pencils are left over? 1 pencil

Number model: 25 ÷ 3 = 8 R1

1 Complete.

in	out
2	4
5	10
10	20
8	16

in

Rule

× 2

out

SRB 74-75

2 Write each number in expanded form.

Example: 579 = 500 + 70 + 9

251 = 200+50+1

425 = 400+20+5

640 = 600+40+0

3 Cross off names that do not belong. Add at least 2 different names.

10	
5 × 2	2 + 2
10 × 10	20 ÷ 2

10×2=20

10÷2=5

4 Solve.

12 = 2 × 6

12 = 6 × 2

5 × 7 = 35

7 × 5 = 35

5 **Favorite Sports**

football ☺☺☺☺☺☺☺☺

soccer ☺☺☺☺☺☺☺

baseball ☺☺☺☺

basketball ☺☺☺☺

Key: ☺ = 1 child

How many more children chose football than baseball as their favorite sport?

4 children

6 Round 91 and 62 to the nearest 10. Use the rounded numbers to estimate. Then solve.

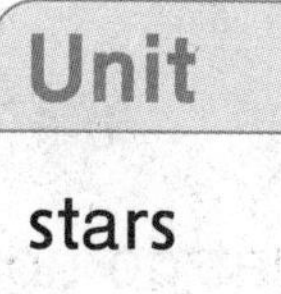

Estimate:

90 – 60 = 30

91 – 62 = 29

Exploring Even and Odd Patterns with Arrays

Lesson 2-10

DATE TIME

Math Message

Amanda wants 2 rows of tomato plants in her garden.
Make sketches to show your thinking.

Can Amanda make an array with 2 equal rows if she has:

9 tomato plants? X

12 tomato plants? ✓

14 tomato plants? ✓

15 tomato plants? X

What do you notice about the numbers of plants that could be planted in arrays with 2 equal rows? What do you notice about the numbers of plants that could *not* be planted in arrays with 2 equal rows?

More Multistep Number Stories

Lesson 2-10

DATE TIM

Use the Dollar Store Poster on *Student Reference Book*, page 270 and solve each problem. Show your work and write number models to keep track of your thinking.

1. Ms. Martin buys 5 packages of brilliant color markers. She then buys 1 package of scented markers. How many markers does she have in all?

Answer: 72 markers

Number models: 5×12=60 1×12=12

60+12=72

2. Mrs. Hickson buys 10 packages of "fashion" pens. She gives 12 pens to her classroom helper. How many pens does Mrs. Hickson have left?

Answer: 108 pens

Number models: 10×12=120 120−12=108

Try This

3. Mr. Wilson buys 1 package of 9-inch balloons, 4 packages of party hats, and 3 packages of party horns. How much money does Mr. Wilson spend in all?

Answer: $ 9

Number models: 4+3=7 7+2=9

Math Boxes

Lesson 2-10

DATE TIME

Math Boxes

1 Joe has 5 packages of crackers. Each package has 6 crackers. How many crackers are there in all?

5 × 6 = ?
(number model with ?)

You may draw an array or a picture.

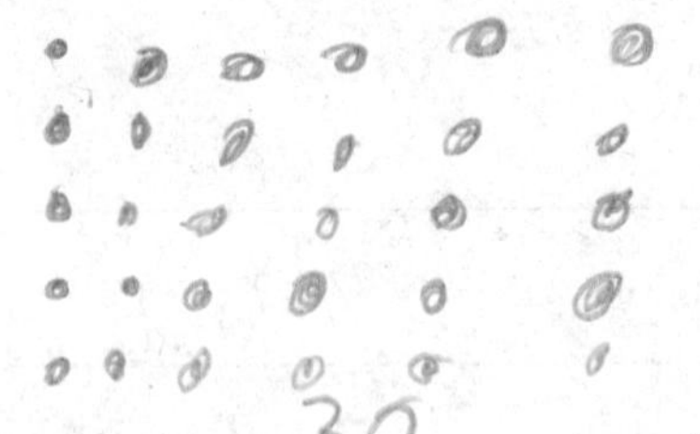

Answer: 30 crackers

2 Sonya had 43 crayons and her teacher gave her 20 more. Now she has 25 more crayons than Mia. How many crayons does Mia have?

Fill in the circle next to the correct answer.

Ⓐ 88

Ⓑ 63

Ⓒ 58

● Ⓓ 38

SRB 30-31

3 Four children share 3 crackers. Use the rectangles to show how they can share the crackers equally.

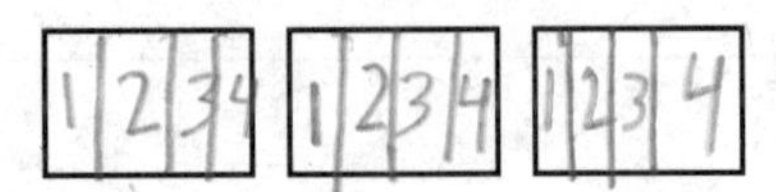

4 Use your Pattern-Block Template. Trace a quadrilateral with 4 right angles.

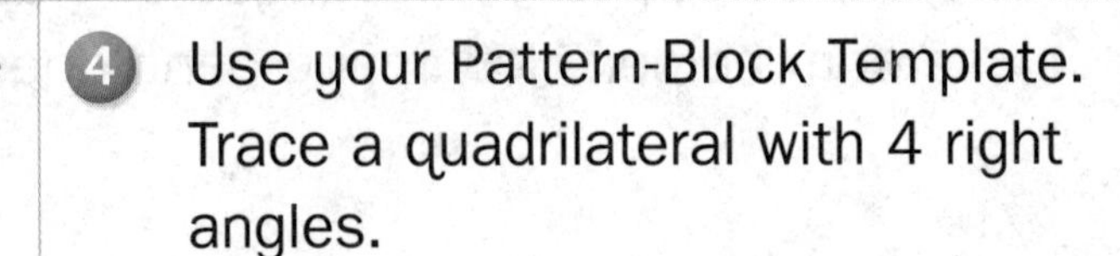

5 **Writing/Reasoning** Look at Problem 3. Write a fraction to name the crackers in each child's share.

You can say three-fourths or $\frac{3}{4}$.

Frames and Arrows

DATE TIME

Find the pattern. Fill in the missing numbers and rule, if needed.

1 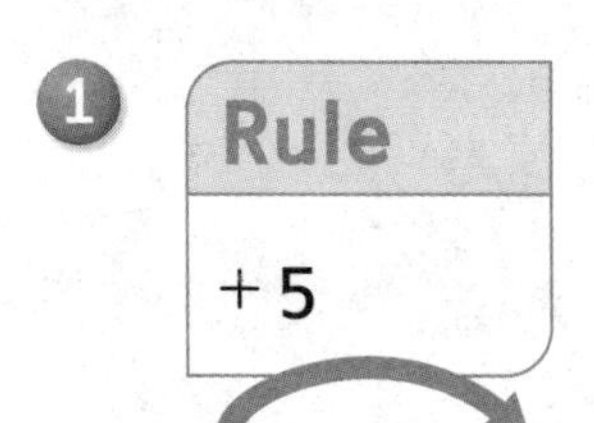

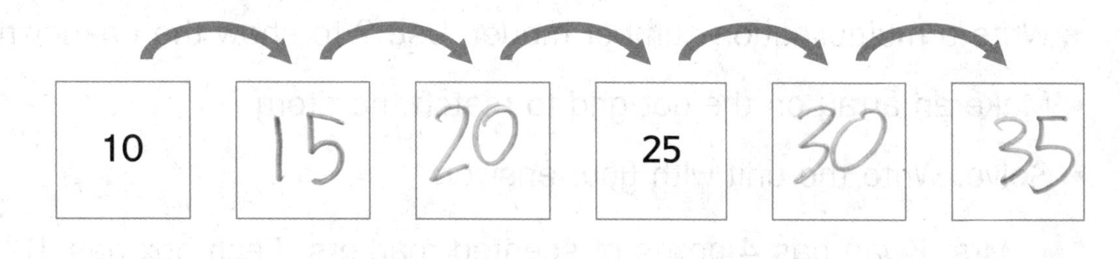

2 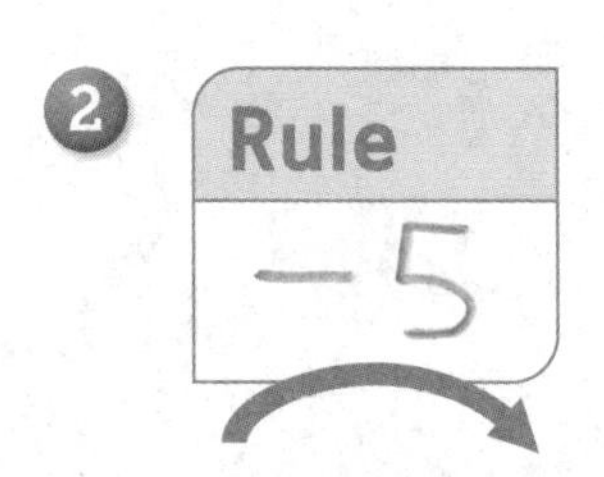

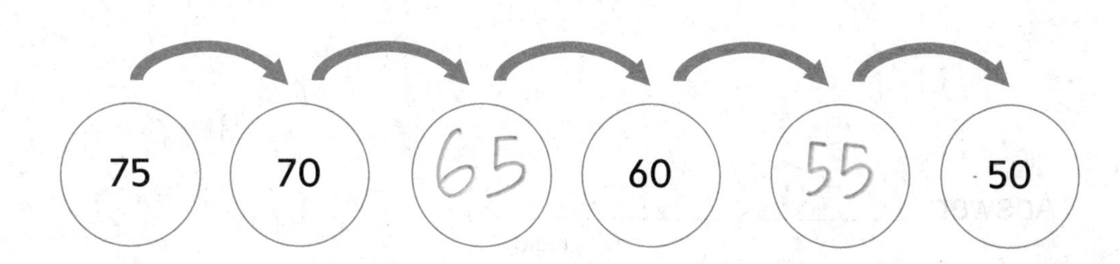

3 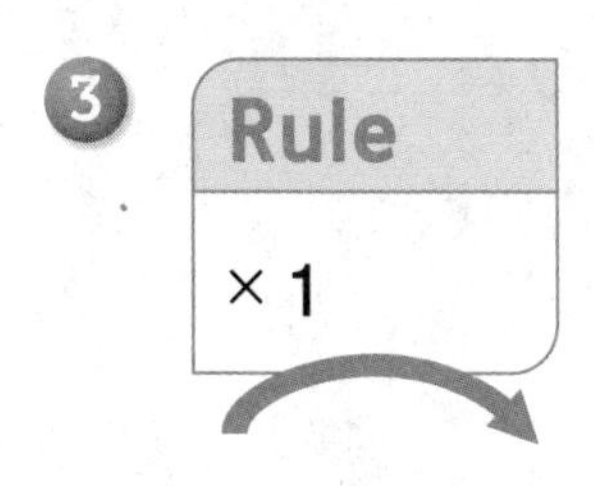

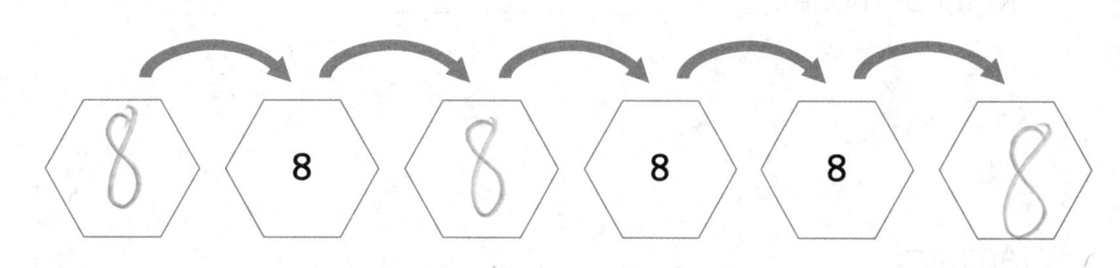

4 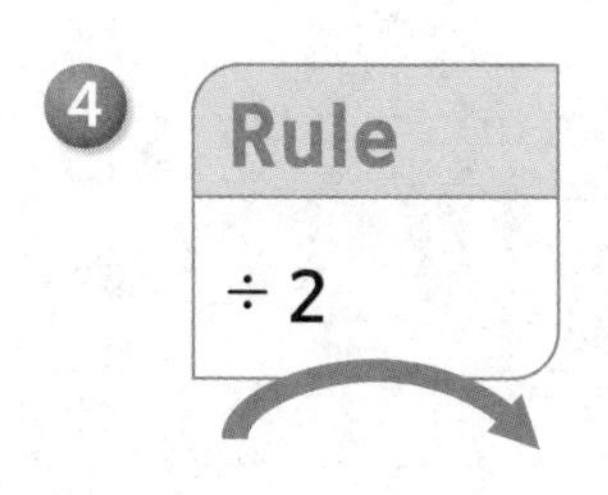

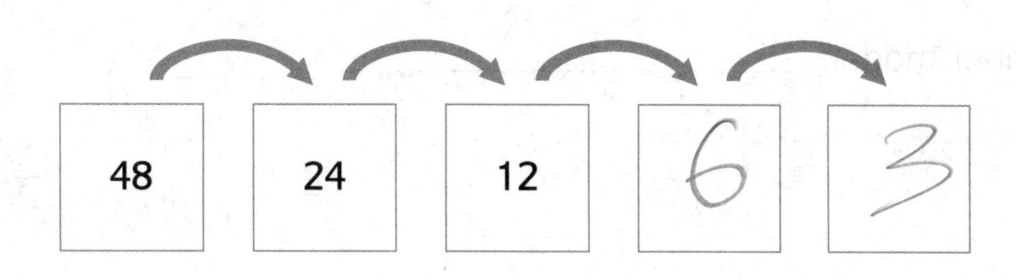

More Number Stories

Lesson 2-11

DATE TIME

For each number story:

- Write a multiplication number model. Use ? to show the unknown.
- Make an array on the dot grid to match the story.
- Solve. Write the unit with your answer.

1. Mrs. Kwan has 4 boxes of scented markers. Each box has 10 markers. How many markers does she have?

Number model: 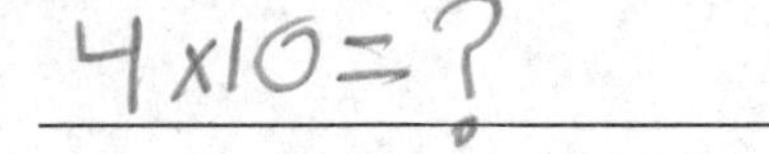

Answer: 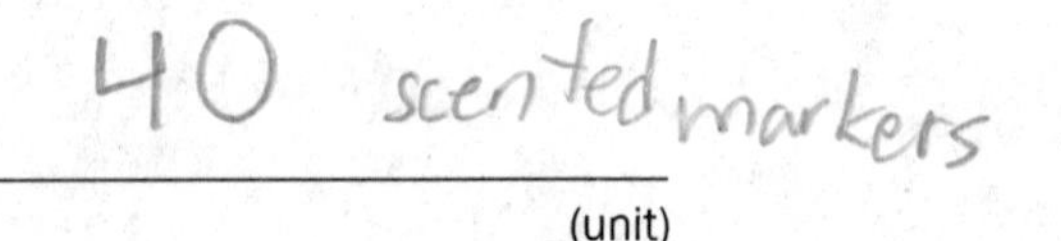

(unit)

2. Monica keeps her doll collection in a case with 5 shelves. On each shelf there are 7 dolls. How many dolls are in Monica's collection?

Number model: 5x7=?

Answer: 35 dolls
(unit)

3. During the summer, Jack mows lawns. He can mow 5 lawns per day. How many lawns can he mow in 9 days?

Number model: 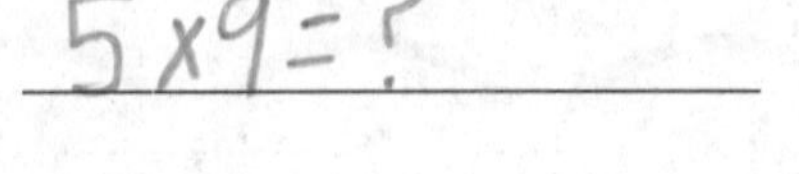

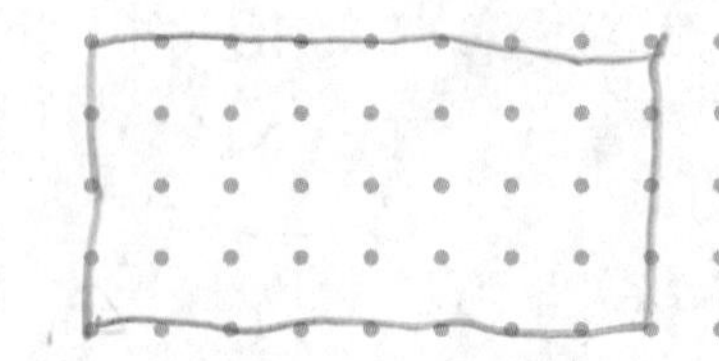

Answer: 45 lawns
(unit)

Math Boxes

Lesson 2-11

DATE TIME

1 Third grade has 81 children. Second grade has 59 children. How many more children are in third grade? You may draw a diagram or picture.

81 − 59 = ?

(number model with ?)

Start 81, Change 22, End 59

Answer: 22 children

2 Olivia has 3 liters of water. One liter of water has a mass of 1,000 grams. What is the mass of 3 liters of water?

1,000 × 3 = 3,000

About 3,000 grams

3 Selene rolls a 10 and draws a 6 card in *Multiplication Draw.* Edgar rolls a 5 and draws a 9 card. Who has the smaller product?

Edgar

Write multiplication number sentences to record their turns.

S. 10×6=60

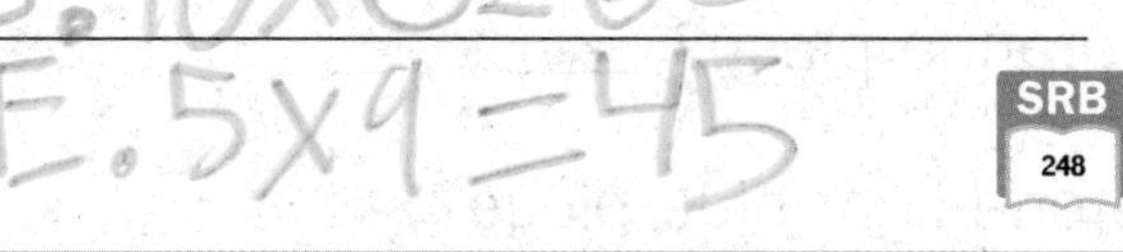

E. 5×9=45

4 Multiply.

4 = 2 × 2

2 × 9 = 18

5 × 2 = 10

35 = 7 × 5

$10 \times 4 = 40$

$10 \times 9 = 90$

5 **Writing/Reasoning** How did you solve Problem 2?

I knowed that in 1 liter there is 1,000 grams. There is

...oration A: ...tion Circles

Lesson 2-12

DATE TIME

Use your fraction circles to answer the questions.

The red circle is the whole.

1. How many yellow pieces cover the red circle? __________
2. How many dark blue pieces cover the red circle? __________
3. How many pink pieces cover the red circle? __________

What fraction or part of the red circle is one pink piece?

____

The pink piece is the whole.

4. How many yellow pieces cover one pink piece? __________
5. How many light blue pieces cover one pink piece? __________

What fraction or part of the pink piece is one light blue piece?

____

The orange piece is the whole.

6. How many light blue pieces cover one orange piece? __________

What fraction or part of the orange piece is one light blue piece?

____

The yellow piece is the whole.

7. How many dark blue pieces cover one yellow piece? __________

What fraction or part of the yellow piece is one dark blue piece?

____

Exploration B: Measuring Area

DATE TIME

Follow the directions on Activity Card 32.

1. a. I traced *the Everything Math Deck*.
   b. It has an area of about ________ square centimeters.
   c. It has an area of about ________ square inches.

2. a. I traced ______________________________.
   b. It has an area of about ________ square centimeters.
   c. It has an area of about ________ square inches.

3. a. I traced ______________________________.
   b. It has an area of about ________ square centimeters.
   c. It has an area of about ________ square inches.

4. a. I traced ______________________________.
   b. It has an area of about ________ square centimeters.
   c. It has an area of about ________ square inches.

Compare your square centimeters and square inches measurements. What do you notice?

____

____

____

Exploration C: Comparing Liquid Volumes

Lesson 2-12

DATE TIME

- Draw containers A, B, and C.
- Circle the container in the top row that you think will hold the most water.
- Below each drawing, show the liquid volume that your container can hold by shading in the 1-liter beaker.

Container A	Container B	Container C
1 liter 1000 ml, 900, 800, 700, 600, 1-half liter 500, 400, 300, 200, 100	1 liter 1000 ml, 900, 800, 700, 600, 1-half liter 500, 400, 300, 200, 100	1 liter 1000 ml, 900, 800, 700, 600, 1-half liter 500, 400, 300, 200, 100

Which container holds the most water? ______________________

Write at least two things you notice about the different liquid volumes.

____

____

____

Math Boxes

DATE TIME

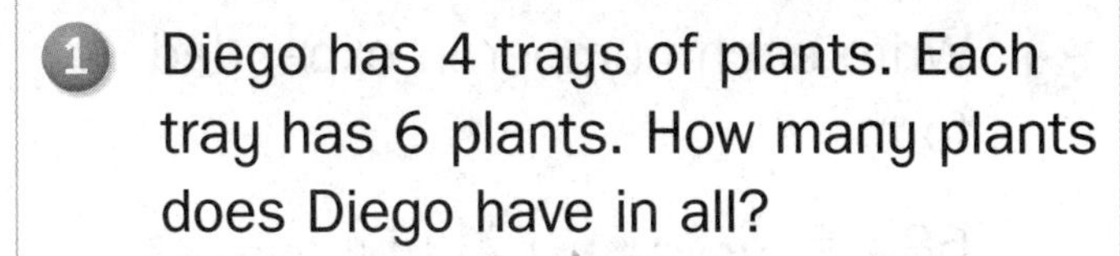

1 Diego has 4 trays of plants. Each tray has 6 plants. How many plants does Diego have in all?

Answer: 24 plants

Fill in the circle next to the correct number model(s).

(A) $6 + 6 + 6 + 6 = 24$

(B) $6 + 4 + 10 = 20$

(C) $4 \times 6 = 24$

(D) $4 + 6 = 10$

2 You have 12 party favors. You put 2 favors in each of 4 party bags. How many party favors are left over?

Number models:

$4 \times 2 = 8$

$12 - 8 = 4$

Answer: 4 party favors (unit)

3 2 friends share 3 oranges. How many oranges will each friend get? Show how they can share the oranges equally.

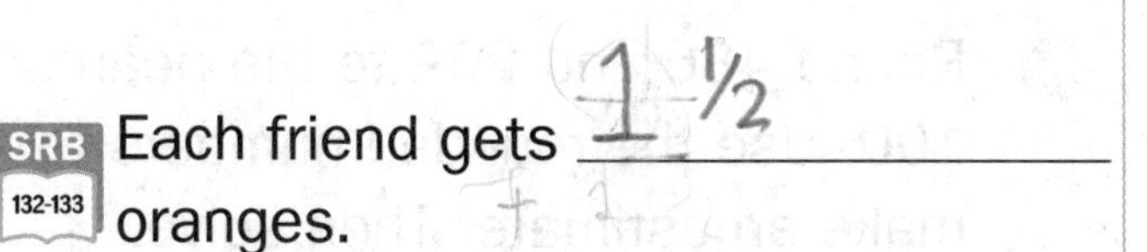

SRB 132-133 Each friend gets 1½ oranges.

4 Use your Pattern-Block Template. Trace a quadrilateral with 2 pairs of parallel sides.

5 **Writing/Reasoning** Explain how you solved Problem 2.

Math Boxes Preview for Unit 3

DATE TIME

1 Complete.

in	out
2	10
4	20
5	25
10	50

in
Rule
× 5
out

SRB 74-75

2 Write each number in expanded form.

684 = 600+80+4

357 = 300+50+7

409 = 400+00+9

890 = 800+90+0

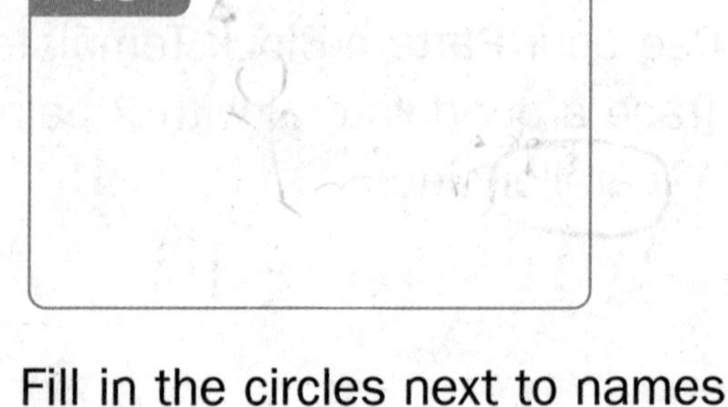

Fill in the circles next to names for 15.

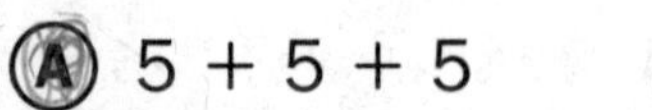

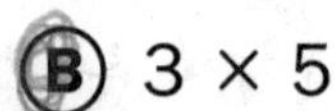

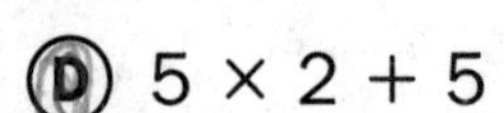

Ⓐ 5 + 5 + 5 Ⓑ 3 × 5

Ⓒ 10 − 5 Ⓓ 5 × 2 + 5

SRB 96-97

4 Solve.

18 = 2 × 9

18 = 9 × 2

5 × 4 = 20

4 × 5 = 20

5 **Favorite Pets of the Class**

dog ☺☺☺☺☺☺☺☺☺
cat ☺☺☺☺☺☺☺
fish ☺☺☺☺
gerbil ☺☺
KEY: ☺ = 1 child

How many more children chose dogs than fish? 5 children

How many children chose fish or cats as their favorite pet?

11 children

6 Round 486 and 209 to the nearest 100. Use the rounded numbers to make an estimate. Then solve.

Estimate: 500-200=300

486 − 209 = 277

"What's My Rule?"

Lesson 3-1

DATE TIME

Fill in the blanks. In the last row of each table, fill in your own pair of *in* and *out* numbers that fit the rule. For Problem 6, write your own rule and fill in the table.

1 Rule: Subtract 50

in	out
100	50
50	0
70	20
150	100
200	150
550	500

2 Rule: add 9

in	out
14	23
34	43
44	53
64	73
94	103
50	59

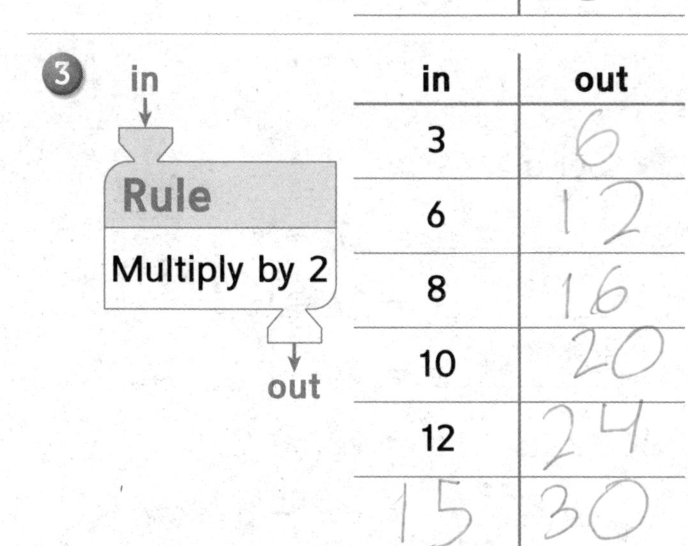

3 Rule: Multiply by 2

in	out
3	6
6	12
8	16
10	20
12	24
15	30

4 Rule: Subtract 30

in	out
60	30
80	50
130	100
230	200
30	0
50	20

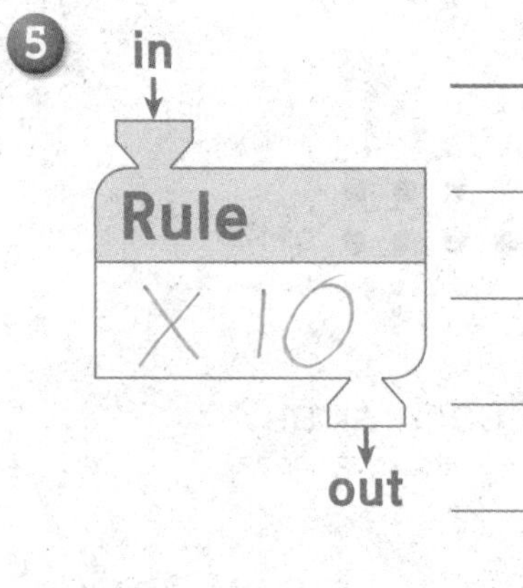

5 Rule: X 10

in	out
2	20
7	70
3	30
6	60
9	90
10	100

6 Rule: ÷ 2

in	out
10	5
100	50
60	30
110	55
20	10
90	45

Math Boxes

Lesson 3-1

DATE TIME

1 Three children rode their bikes a total of 23 miles. One child rode 9 miles and one rode 8 miles. How many miles did the third child ride? Write number models.

Number models: 9+8=17

23-17=6

Answer: 6 miles (unit)

2 Draw an array of 8 circles in 2 equal rows.

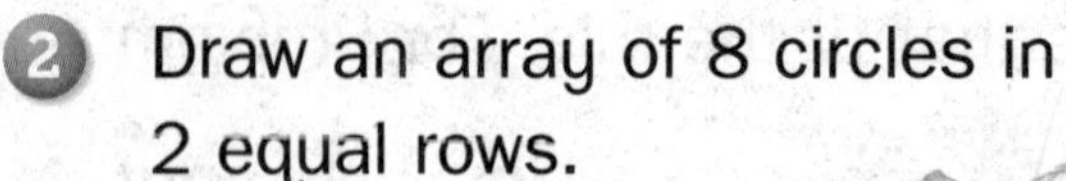

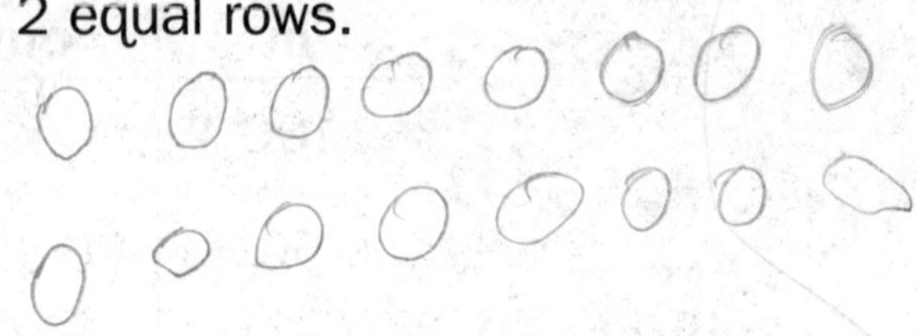

Draw an array of 10 circles in 2 equal rows.

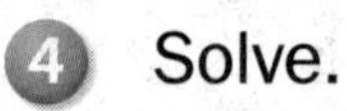

3 Equally share 16 pennies between 2 children. You may draw a picture.

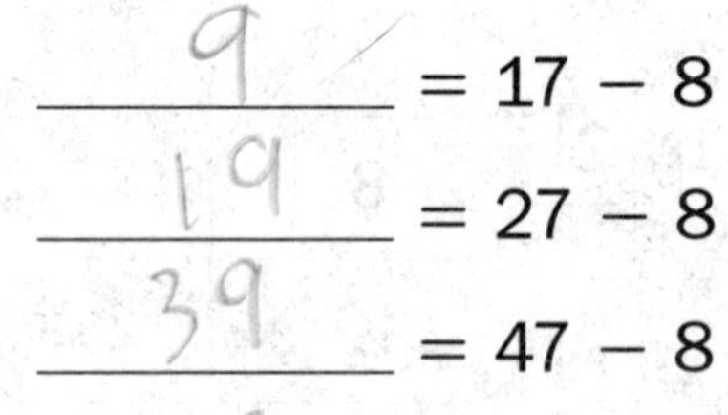

Each child gets 8 pennies.

There are 0 pennies left over.

Number model: 16÷2=8

4 Solve.

Unit
trees

9 = 17 − 8

19 = 27 − 8

39 = 47 − 8

119 = 127 − 8

159 = 167 − 8

SRB 114

5 **Writing/Reasoning** Jenna drew this array.

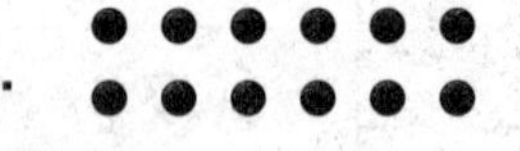

Look at Jenna's array and your arrays in Problem 2.

What do you notice about the total number of circles in each array?

Strategies for Estimation

Lesson 3-2

DATE TIME

Rosa makes an estimate for the addition problem below. She uses numbers that are **close** to the numbers in the problem but are **easier** to use.

322 + 487 = ?

Unit
$

320 + 490 = 810
She rounded to the nearest tens.

1. Explain Rosa's thinking to a partner.
2. Make a different estimate. What **close-but-easier** numbers could you use? Write a number sentence in the thought bubble to show your thinking.

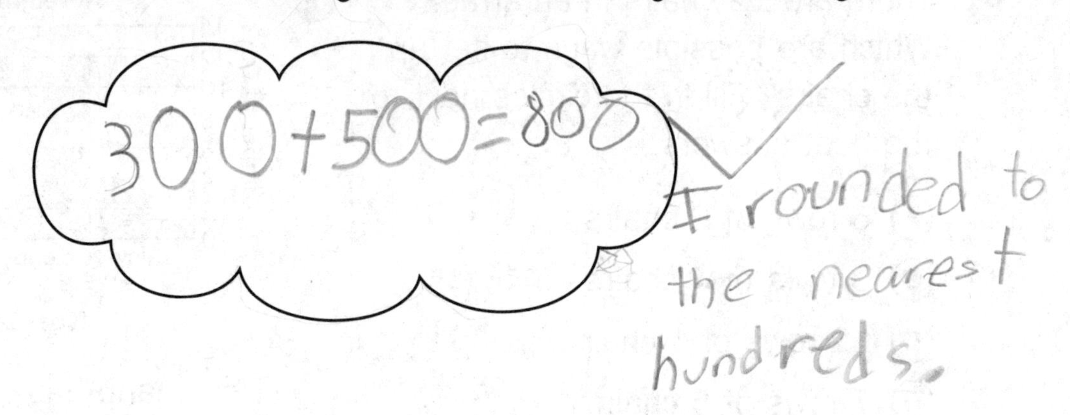

300+500=800
I rounded to the nearest hundreds.

Math Boxes

DATE TIME

1 Jocelyn shares a fruit strip equally among herself and 3 friends. How much does each child get? Use the rectangle below to show the equal shares.

1	2	3	4

Each child gets $\frac{1}{4}$ of a fruit strip.

SRB 132-133

2 You have exactly enough money to buy two grape juice boxes for 45 cents each.

How much money do you have? You may draw a diagram or a picture.

____
(number model with ?)

Answer: ______________ (unit)

3 There are 12 chairs in an array. Which are possible ways to set up the chairs? Fill in the circles next to the best answers.

- (A) 6 rows of 2 chairs
- (B) 3 rows of 4 chairs
- (C) 12 rows of 1 chair
- (D) 7 rows of 5 chairs

SRB 41-43

4

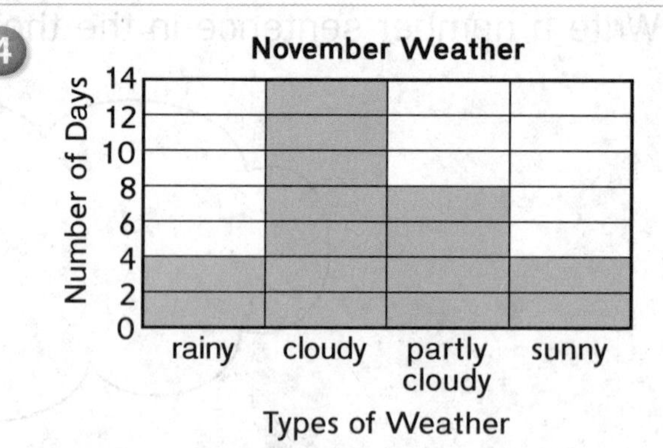

How many days were not rainy?

26 days
(unit)

5 Fill in the empty frames.

Rule: + 25

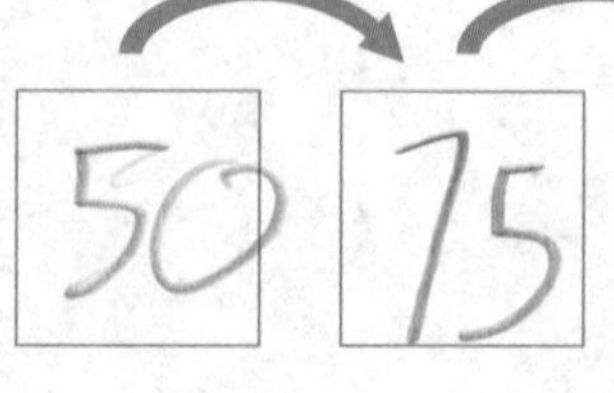

100 | 125 | 150

SRB 72-73

Partial-Sums Addition

Lesson 3-3

DATE TIME

Unit

Estimate. Write number sentences to show how you estimated. Use partial-sums addition to solve Problems 1 and 2. Use any strategy to solve Problem 3. For all problems, show your work and check that your answers make sense.

Example: 329 + 418 = 747

Estimate: 300 + 400 = 700

```
                  329
                + 418
300 + 400  →      700
 20 + 10   →       30
  9 + 8    →       17
                  747
```

1 143 + 28 = 171

Estimate: 140+30=170

```
  143
+  28
  100
   60
   11
  171
```

2 195 + 537 = 732

Estimate: 200+500=700

```
  195
+ 537
  600
  120
   12
  732
```

3 378 + 439 = 817

Estimate: 400+400=800

```
  378
+ 439
  700
  100
   17
  817
```

Math Boxes

Lesson 3-3

DATE TIME

Math Boxes

1 Luisa has 5 bags that each hold 5 peaches. She drops 6 peaches. How many peaches does she have now?

Write number models to show your steps.

Number models: $5 \times 5 = 25$

$25 - 6 = 19$

Answer: 19 peaches (unit)

SRB 30-31

2 Circle the number of Xs that you can draw in an array with 2 equal rows.

19 14

Draw that number of Xs in an array with 2 equal rows.

X X X X X X X
X X X X X X X

Can you make an array with 2 *equal* rows with an odd number of Xs?

no

3 14 books are equally shared among 3 children. You may draw a picture.

How many books does each child get? 5 (unit)

How many books are left over? 1 (unit)

Number model: ______________

SRB 39-40

4 Solve.

Unit
soccer balls

______ = 15 – 7

______ = 35 – 7

______ = 85 – 7

______ = 135 – 7

______ = 235 – 7

SRB 114

5 **Writing/Reasoning** Explain how you can use the basic fact 15 – 7 to help solve the other number sentences in Problem 4.

SRB 114

Column Addition

Lesson 3-4

DATE TIME

Unit

Estimate. Then use column addition to solve Problems 1 and 2. Use any strategy to solve Problem 3. Use your estimates to check whether your answers make sense.

Example: 148 + 59 = ?

Estimate: *150 + 60 = 210*

100s	10s	1s
1	4	8
+	5	9
1	9	17
1	10	7
2	0	7

148 + 59 = *207*

1 67 + 25 = ?

Estimate: 70+30=100

	6	7
+	2	5
	8	12
	9	2

67 + 25 = ______

2 227 + 386 = ?

Estimate: 230+390=620

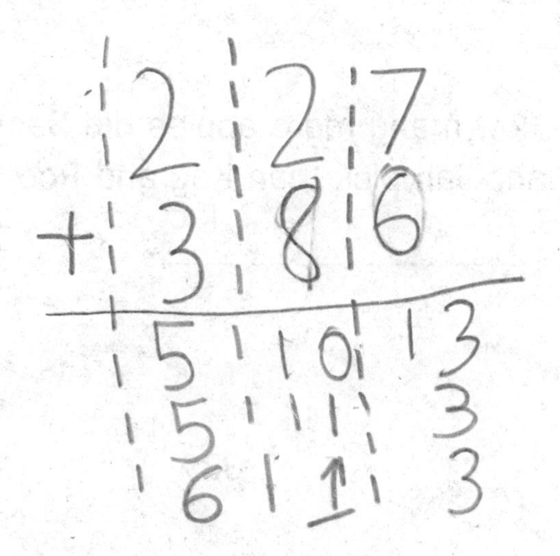

227 + 386 = ______

3 481 + 239 = ?

Estimate: 480+240=720

4	8	1
+ 2	3	9
6	11	10
7	12	0
7	2	0

481 + 239 = 720

Math Boxes

1. Fill in the circles next to the pictures that show 2 plums shared equally by 3 children.

Ⓐ 

Ⓑ 

Ⓒ 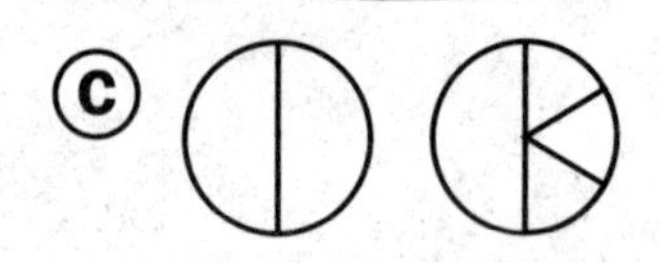

Ⓓ 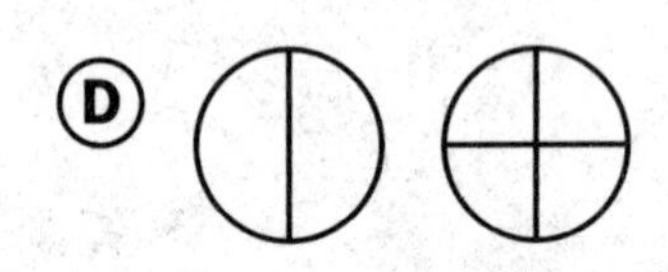

SRB 132-133

2. Dontrell gives his brother 65¢ to buy orange juice. Now his brother has 90¢. How much money did his brother have to start?

90 − 65 = ?
(number model with ?)

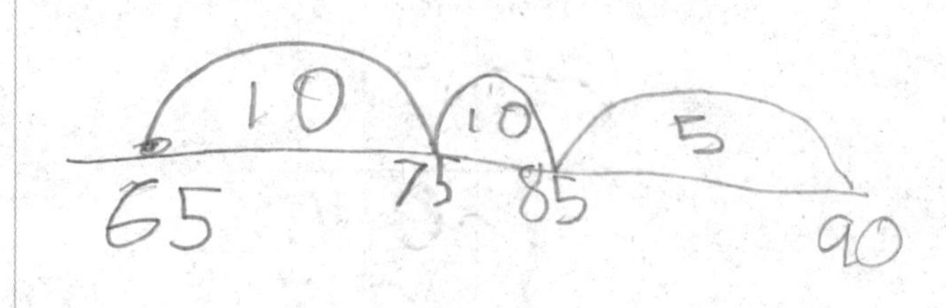

Answer: ¢25
(unit)

SRB 76

3. Arrange 12 stars in an array. Write a multiplication number model that fits your array.

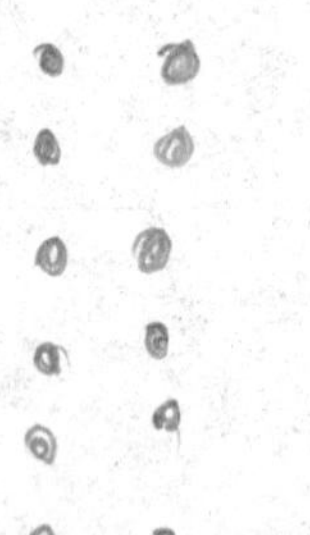

Number model: 6 x 2 = 12

SRB 41-42

4. 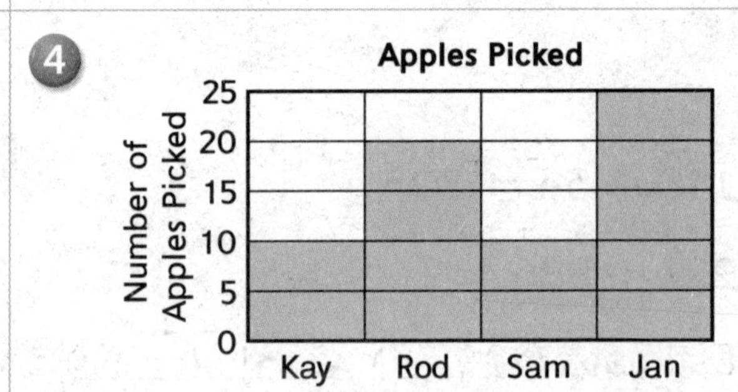

How many more apples did Sam and Jan pick than Kay and Rod?

5 apples
(unit)

5. 

SRB 72-73

Counting-Up Subtraction

Lesson 3-5

DATE TIME

Fill in the unit box. For each problem, make an estimate. Count up to solve Problems 1–3. Use open number lines or number sentences. Use any strategy to solve Problem 4. Show your work. Use your estimates to check whether your answers make sense.

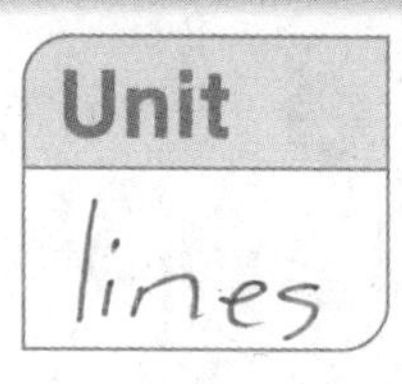

1. 67 – 37 = ?

Estimate: 70 - 40 = 30

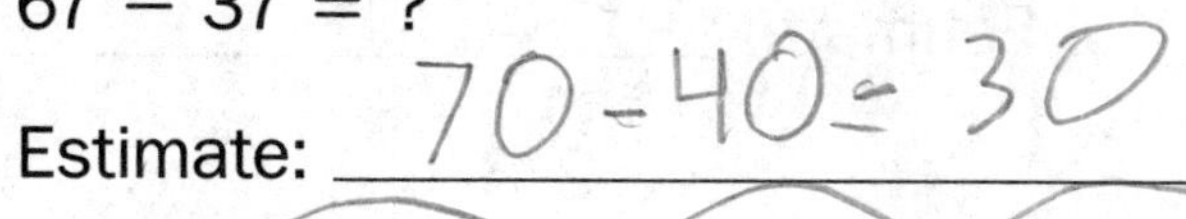

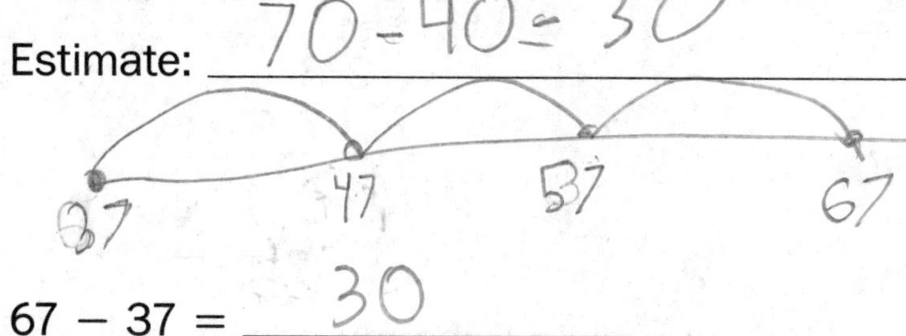

67 – 37 = 30

2. ? = 91 – 46

Estimate: 40 = 90 - 50

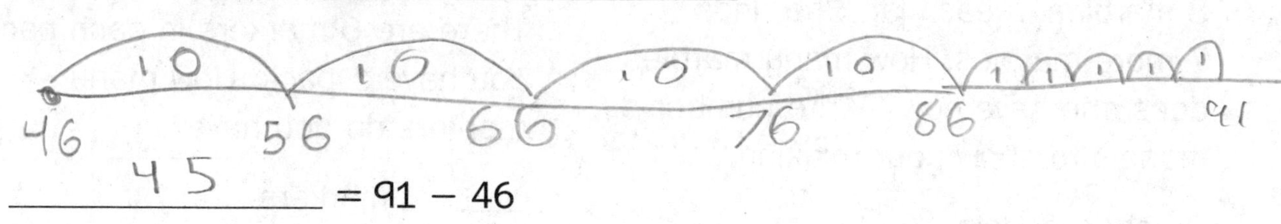

45 = 91 – 46

3. ? = 283 – 256

Estimate: 300 - 300 = 0

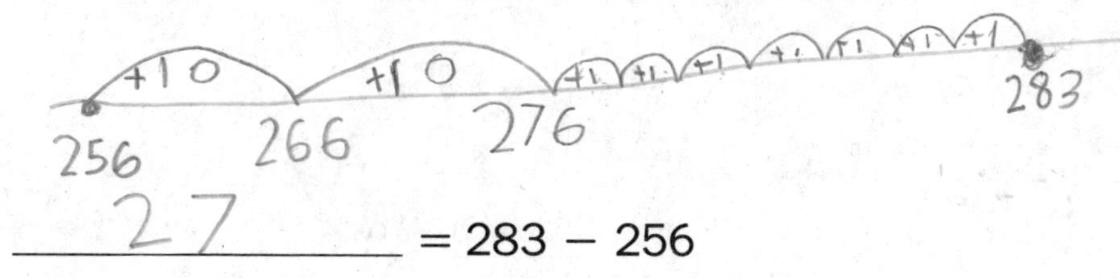

27 = 283 – 256

4. 752 – 487 = ?

Estimate: 800 - 500 = 300

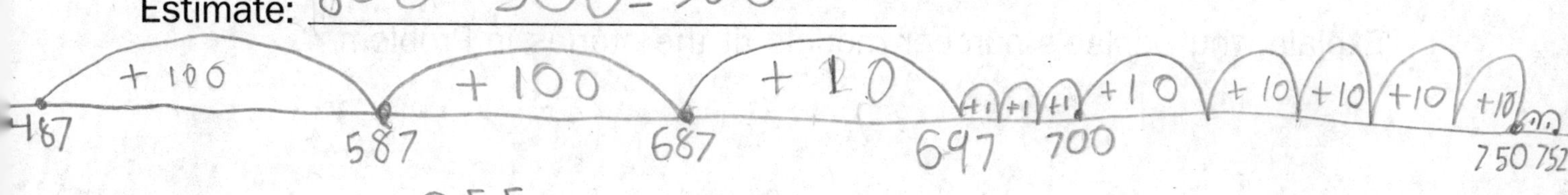

752 – 487 = 355

Math Boxes

Math Boxes

1 Fill in the missing numbers.

in

Rule
× 2

out

in	out
5	10
7	14
8	16
10	20
100	200

SRB 74-75

2 Round each addend to the nearest 100 to make an estimate. Then solve.

Unit

Estimate: 200 + 100 = 300

$$\begin{array}{r} 236 \\ +\ \ 79 \\ \hline \end{array}$$

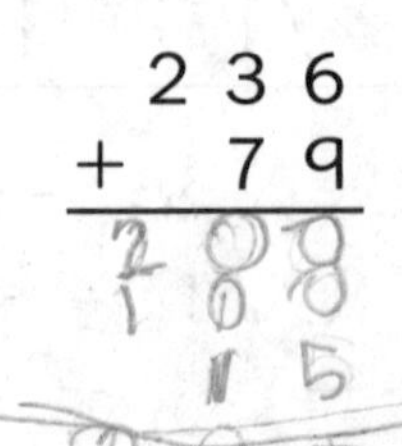

Think: Does my answer make sense?

SRB 106, 116-118

3 Stacy has 2 jars of marbles with 8 marbles in each jar. She finds 7 more marbles. How many marbles does she have now? Write number models to show your thinking.

Number models: 2×8=16
16+7=23

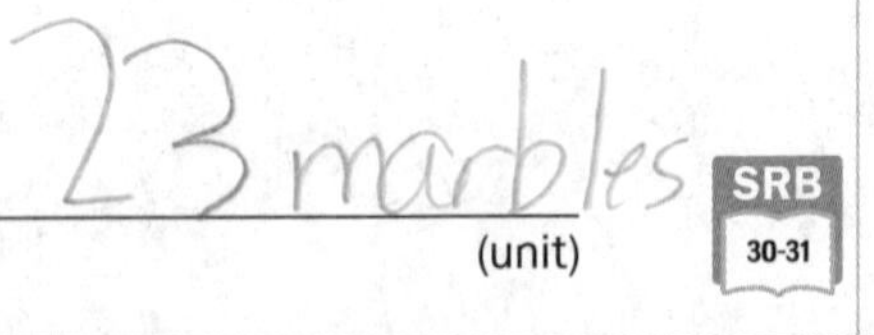

Answer: 23 marbles (unit)

SRB 30-31

4 Solve.

There are 6 markers in each pack. You have 1 pack. How many markers do you have?

6 markers

There are 96 eggs in one green turtle clutch. How many eggs are there?

96 eggs

SRB 46

5 **Writing/Reasoning** Denise wrote:

$1 \times 6 = 6$ $1 \times 96 = 96$

Explain why Denise's number models fit the stories in Problem 4.

She did it correct because there is 1 pack of mar

SRB 46

Expand-and-Trade Subtraction

Lesson 3-6

DATE TIME

Fill in the unit box. For each problem, write a number sentence for your estimate. Write each number in expanded form. Solve using expand-and-trade subtraction. Compare your answer to your estimate. Does your answer make sense?

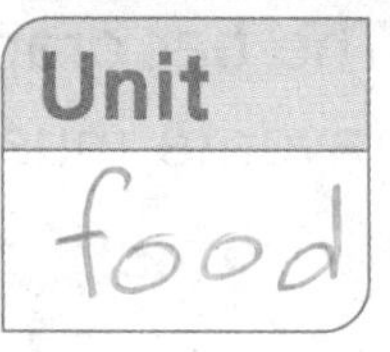

Example: 247 – 186 = ?

Estimate: *250 – 200 = 50*

```
          100  140
247 → 200 + 40 + 7
– 186 → 100 + 80 + 6
             60 + 1 = 61
```

247 – 186 = *61*

1. 65 – 47 = ?

Estimate: 70-50= 20

```
           50 15
65 ——→ 60 + 5
– 47 ——→ 40 + 7
18       10 + 8
```

65 – 47 = 18

2. 182 – 56 = ?

Estimate: 180+60= 120

```
               70 12
182 ——→ 100 + 80 + 2
– 56 –        50 + 6
126     100 + 20 + 6
```

182 – 56 = 126

3. 341 – 225 = ?

Estimate: 340-230=110

```
              30 11
341 ——→ 300 + 40 + 1
– 225 —→ 200 + 20 + 5
116      100 + 10 + 6
```

341 – 225 = 116

Comparing Data in a Bar Graph

Lesson 3-6

DATE TIME

The bar graph shows the number of children in Grade 3 who chose each type of favorite music. Use the graph to solve the number stories.

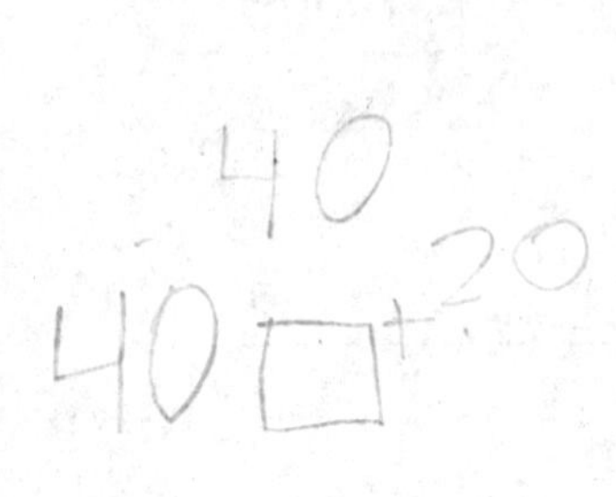

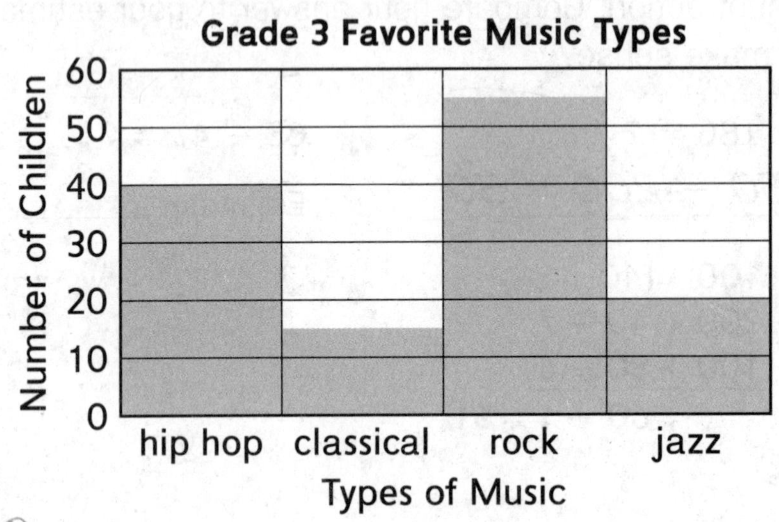

1. How many more children like rock than jazz? ______ children

2. How many fewer children like classical than hip hop? ______ children

3. How many more children like hip hop than classical and jazz together?

   ______ children

4. Write a number story that could be solved using the graph. Write the answer to your story.

   ____

   ____

   ____

   Answer: ______________

Math Boxes

DATE TIME

1 The normal spring high temperature in Los Angeles is 72°F. The normal low is 56°F. What is the difference between the temperatures?

Number model: 72 − 56 = 16

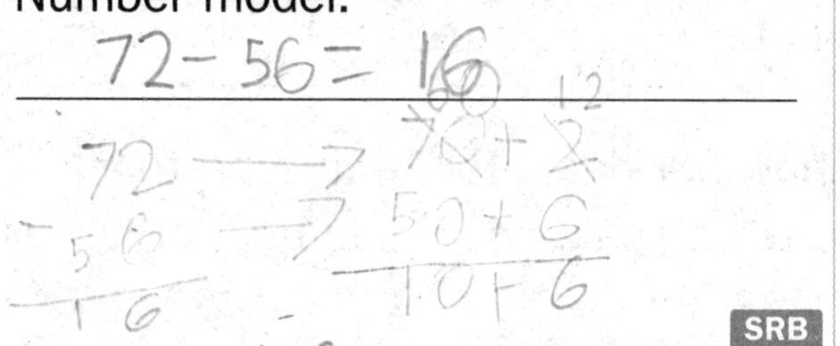

Answer: 16 °F

SRB 76

2 Fill in the circles next to the measurements that are about equal to the mass of 1,000 paper clips. (*Hint:* 1 paper clip is about 1 gram.)

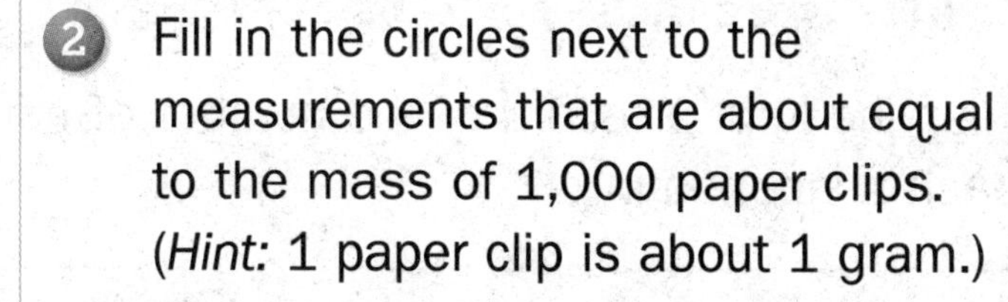

Ⓐ about 10 grams

Ⓑ about 1,000 grams

Ⓒ about 1 kilogram

Ⓓ about 100 kilograms

SRB 183

3 Rita has 26 stickers and shares them equally among herself and two friends. How many stickers does each get?

Each child gets 8 stickers.

There are 2 stickers left over.

Number model: 26 ÷ 3 = 8R2

SRB 39-40

4 Complete the Fact Triangle. Write the fact family.

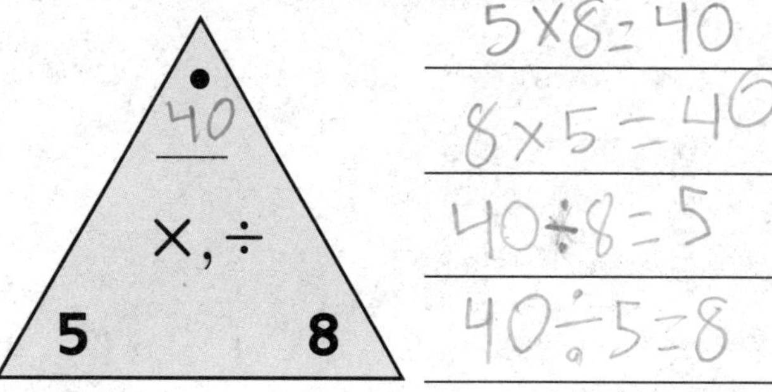

5 × 8 = 40

8 × 5 = 40

40 ÷ 8 = 5

40 ÷ 5 = 8

SRB 53

5

Record the time. 4:35

What time will it be in 25 minutes? 5:00

SRB 18-19, 186

6 A soccer ball has a mass of about 425 grams. A softball has a mass of about 184 grams. What is their total mass?

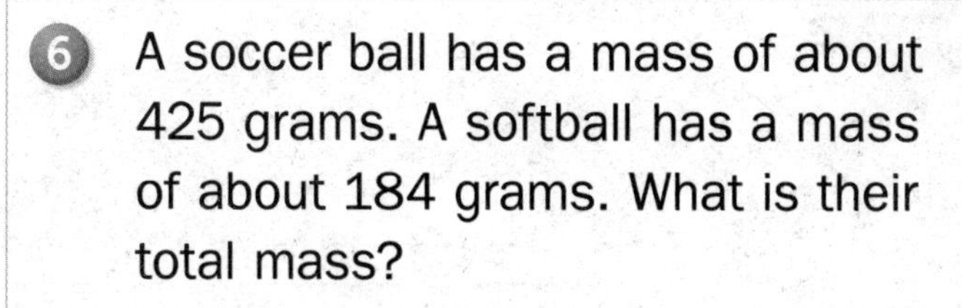

____
(number model with ?)

Answer: about ______________
(unit)

SRB 76

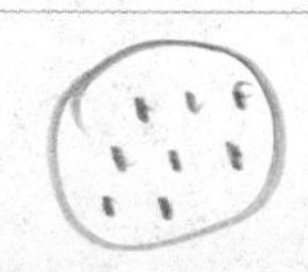

Scale for a Data Set

Lesson 3-7

DATE TIME

Math Message

Jasmine kept a record of the number of minutes she did homework each school day.

Monday	Tuesday	Wednesday	Thursday	Friday
45	20	35	40	17

She wants to use the bar graph below to show her data. Talk with a partner about how Jasmine could set up her graph. What scale could she use for Number of Minutes?

____

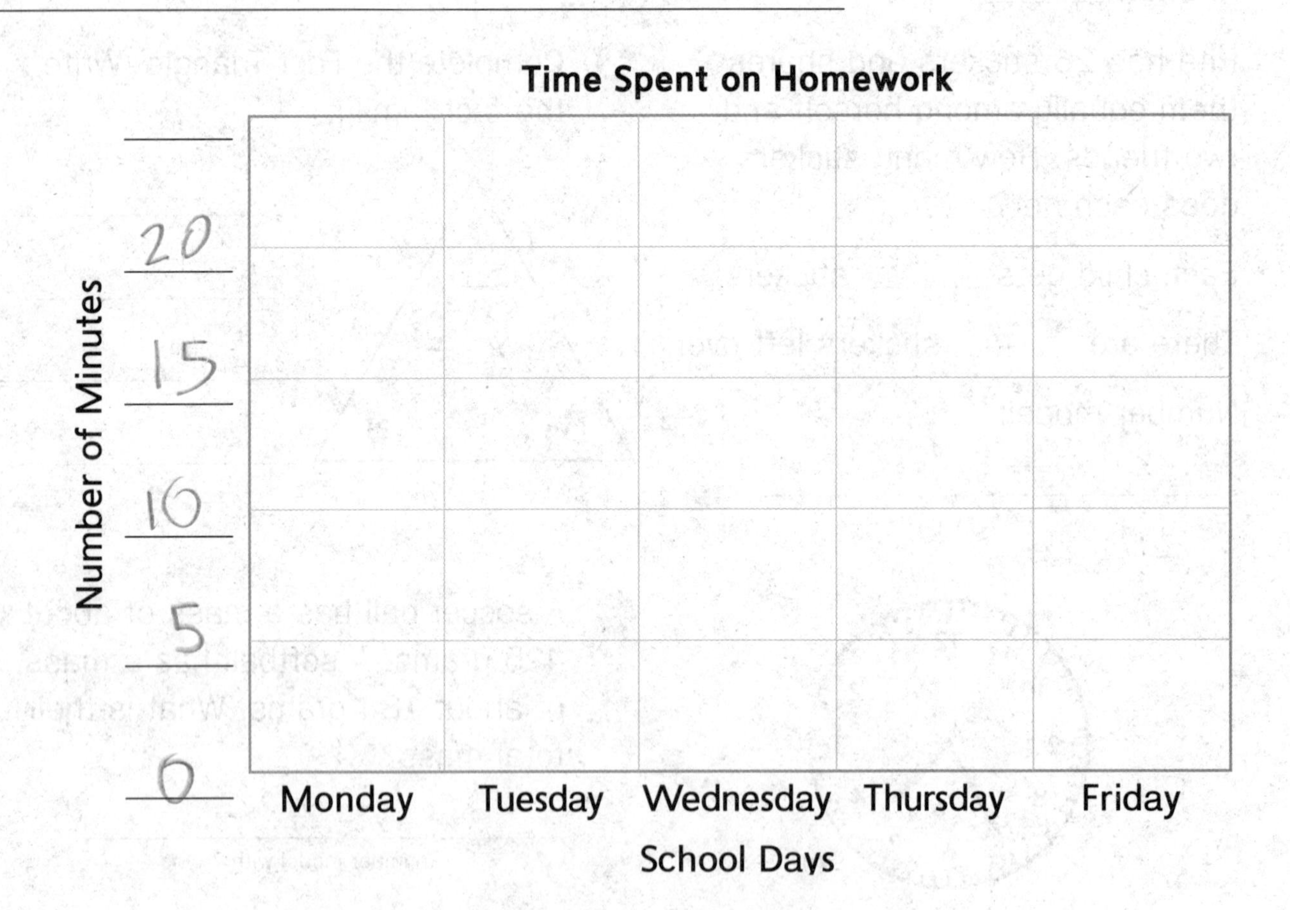

Exploration A: Pattern-Block Sort Bar Graph

Lesson 3-7

DATE TIME

Record the number of pattern blocks in each group.

Triangle	Wide Rhombus	Narrow Rhombus	Hexagon	Trapezoid
卌 ||||	卌 |	||||||||	卌 卌||	卌 卌

Choose a scale for your graph based on your data. Graph your pattern-block data.

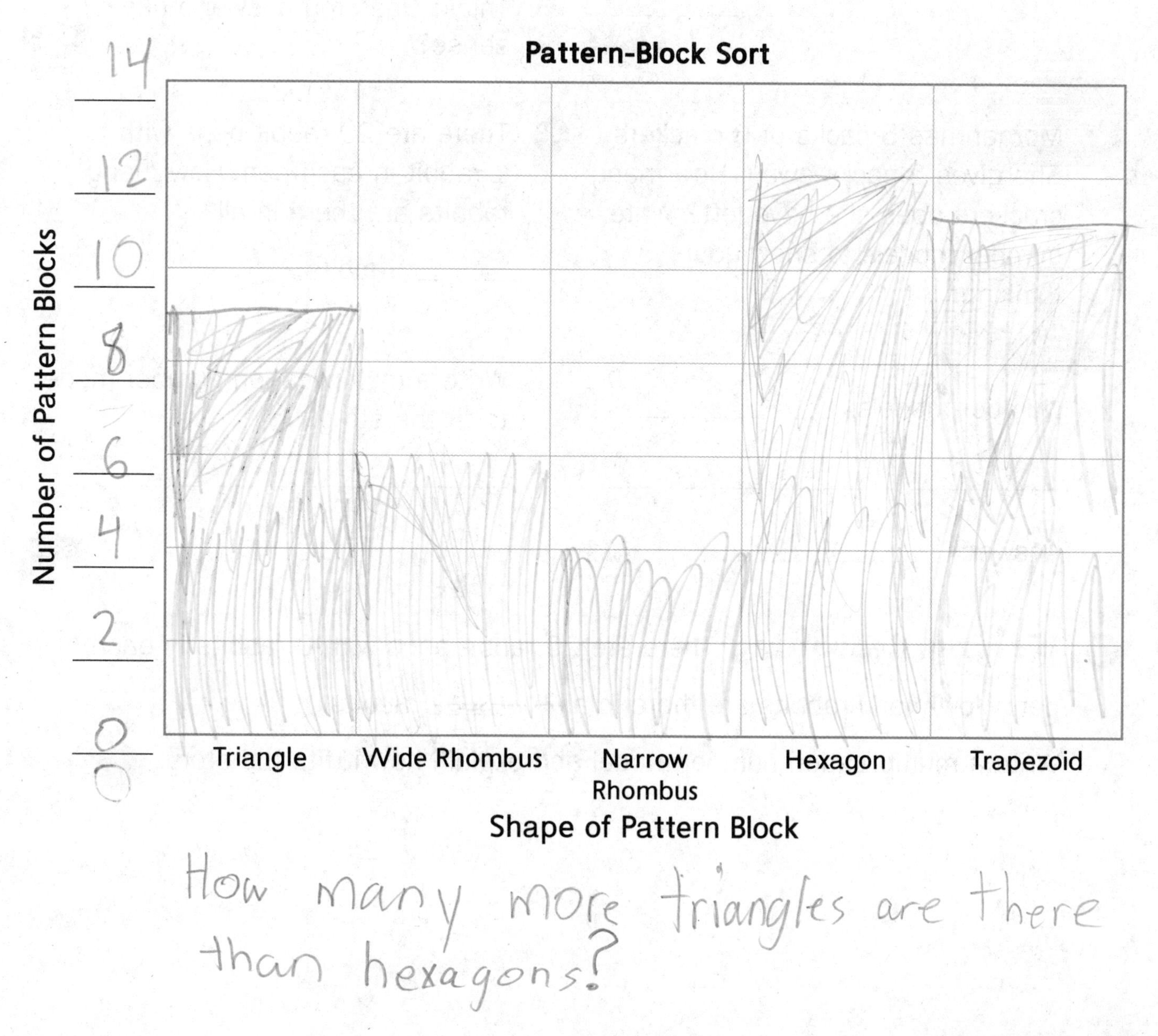

How many more triangles are there than hexagons?

Math Boxes

Lesson 3-7

DATE TIME

Math Boxes

1 Fill in the missing numbers.

in → Rule $\div 2$ → out

in	out
4	2
10	5
18	9
160	50
50	25

SRB 74-75

2 Round each addend to the nearest 10 to make an estimate. Then solve.

Unit: books

Estimate: 470+60=530

$$\begin{array}{r} 472 \\ +\ 59 \\ \hline \end{array}$$

Think: Does my answer make sense?

SRB 106, 116-118

3 Morgan has 5 packs of 6 crackers. She gives 2 packs away. How many crackers does she have left? Write number models to show your thinking.

Number models: 5x6=30

30-12=18

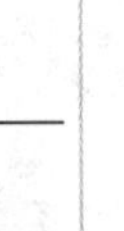

Answer: 18 cracker (unit)

SRB 30-31

4 There are 20 rabbit pens with 1 rabbit in each pen. How many rabbits are there in all?

20 rabbits

Write a multiplication number model to fit the story.

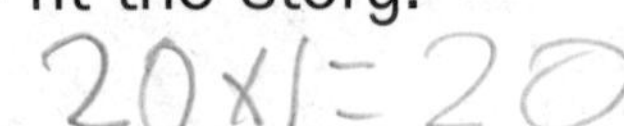

SRB 46

5 **Writing/Reasoning** There are 20 rabbit pens with 0 rabbits in each pen. How many rabbits are there in all? 0 rabbits

Write a multiplication number model and explain how it fits the story.

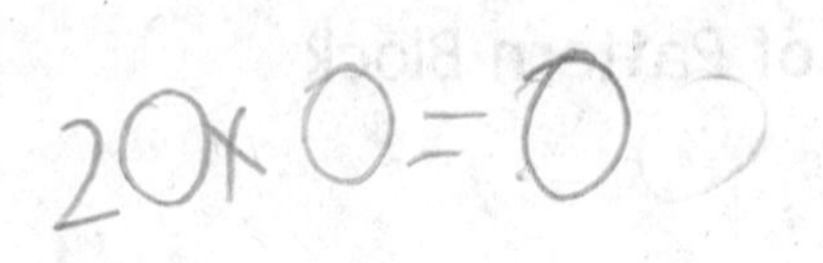

SRB 46

Creating a Scaled Picture Graph

Lesson 3-8

DATE TIME

Math Message

Copy your pattern-block data from journal page 79 into the table below.

Triangle	Wide Rhombus	Narrow Rhombus	Hexagon	Trapezoid
9	6	4	12	10

Title: ______

Triangle

Wide Rhombus

Narrow Rhombus

Hexagon

Trapezoid

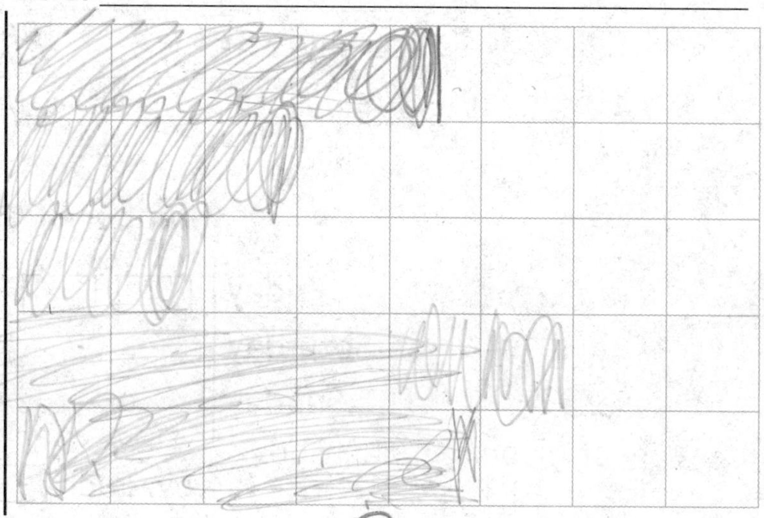

Key: Each □ = 2 pattern blocks

Drawing a Scaled Picture Graph

DATE TIME

Kellogg School held a weekend car wash. Use the data in the tally chart and the key to complete the picture graph below. You may refer to pages 193–194 in your *Student Reference Book*.

Number of Cars Washed	
Day	**Number of Cars**
Friday	𝍸 |||
Saturday	𝍸 𝍸 𝍸 |
Sunday	𝍸 𝍸 ||||

Number of Cars Washed

Friday | ▭ ▭
Saturday |
Sunday |

Key: Each ▭ = 4 cars

1. Why are there 2 rectangles next to Friday?

____

____

2. How did you figure out how many car symbols to draw for Sunday?

____

____

____

Math Boxes

1 The normal spring high temperature in Seattle is 63°F. The normal low is 17 degrees cooler. What is the normal low temperature?

Number model: 63 − 17 = 16

63 → 60 + 13
−17 → 10 + 7
46 40 + 6

Answer: 46 °F

SRB 76

2 Name something in the classroom that has a mass of about 1 kilogram.

table

How do you know?

SRB 183

3 There are 20 chairs. Each table has 5 chairs. How many tables are there? You may draw a picture.

Fill in the circle next to the correct answer.

- (A) 3 tables
- (B) 4 tables
- (C) 6 tables
- (D) 10 tables

SRB 39-40

4 Complete the Fact Triangle. Write the fact family.

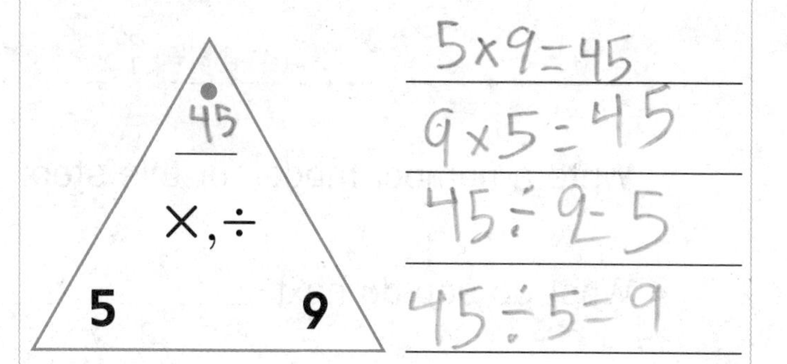

SRB 53

5

What time is it? 3:10

What time will it be in 45 minutes? 3:55

SRB 186

6 A baseball has a mass of about 142 grams. A tennis ball has a mass of about 57 grams. About how many more grams is the baseball than the tennis ball?

142 − 57 = ?
(number model with ?)

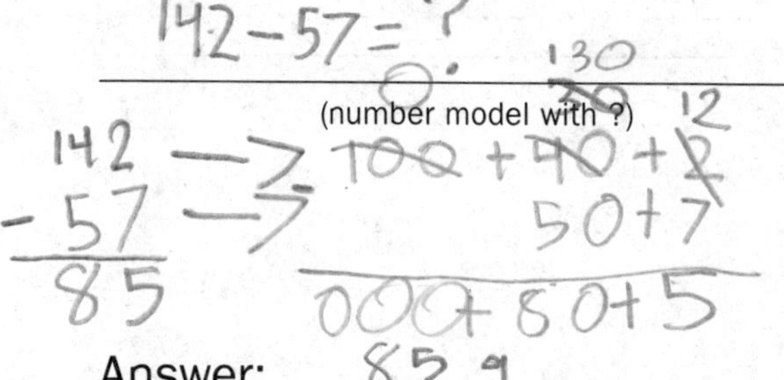

Answer: 85 g
(unit)

SRB 76

Making Sense of Number Stories

Lesson 3-8

DATE TIME

1. There are 4 crackers in each pack. You buy 3 packs and give 1 pack to your friend. How many crackers do you have left?

- What do you know from the problem? 4 crackers = 1 pk.
  Buy 3 pks. gave away 1 pk.

- What do you need to find out? How many crackers left

- What is your plan? I should multiply 4 and 3 together and I should minus 4 to get the answer

- What do you do first? I will multiply 4 and 3 together first.

  Write a number model for this step: 4 x 3 = 12

- What do you do next? I will minus 4 from 12 next.

  Write a number model for the second step: 12 - 4 = 8

  There are 8 crackers left.

- How do you know your answer makes sense? I could make a 4 by 3 array and then cross 4 dots.

Making Sense of Number Stories (continued)

Lesson 3-8

DATE TIME

2. Each pack of crackers costs 30¢. You have $1 (100¢). How much change will you get after you buy 3 packs?

- What do you know from the problem? Each pk = 30¢ $1
- What do you need to find out? How many money left
- What is your plan? ______
- What do you do first? ______

  Write a number model for this step: ______
- What could you do next? ______

  Write a number model for this step: ______

  I have ______¢ left.
- How do you know your answer makes sense? ______

Exploring Arrays with Equal Factors

DATE TIME

Work with a partner.

Materials ☐ Centimeter Grid Paper (*Math Masters,* page TA19)
☐ centimeter cubes
☐ tape

Directions

1. Choose a number 1 through 10. Use centimeter cubes to build an array with that number of rows and that same number of columns.
2. Record the array on centimeter grid paper. Use Xs or color in each square. Write a multiplication number sentence below each array. Example:

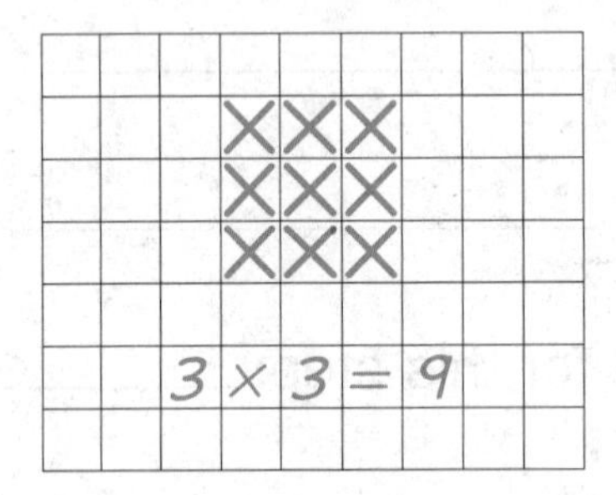

3. Repeat Steps 1 and 2 with at least two other numbers. You may need to tape pieces of grid paper together for the larger arrays.
4. Look at the arrays you made. How are they similar? How are they different?

Math Boxes

DATE TIME

1. Amy arrived at the library at 1:10 P.M. She left at 1:55 P.M. How long was she at the library?

You may use your toolkit clock or draw an open number line.

Answer: 45 min.
(unit)

2. Fill in the missing numbers.

in → Rule ×5 → out

in	out
1	5
5	25
6	30
10	50
100	500

SRB 74-75

3. Round to the nearest 10 to make an estimate. Then solve.

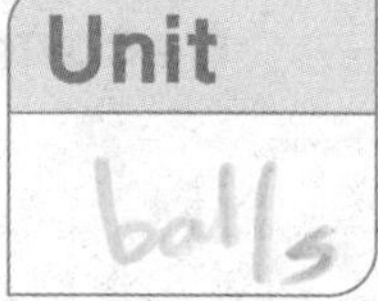

Estimate: 130 - 40 = 90

$$\begin{array}{r} 127 \\ -\ 39 \\ \hline 088 \end{array}$$

Think: Does my answer make sense? Yes

4. Use the dots below to show a 4-by-6 array.

Write a number sentence to match your array.

4 x 6 = 24

5. **Writing/Reasoning** Write a number story to go with the array in Problem 4.

The was 4 boxes. Each box has 6 balls, How many balls do they have in all?

A Multiplication Rule

Lesson 3-10

DATE TIME

Roll a die twice to get 2 factors. Sketch an array using those 2 factors and record a number sentence to match. Switch the factors and record an array and number sentence to match.

Example: I roll a 3 and a 4:

First Array

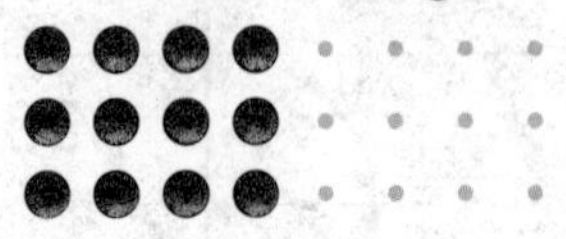

$3 \times 4 = 12$

Second Array

$4 \times 3 = 12$

1. Factors I am using: 6 and 5

First Array

Number sentence: 6x5=30

Second Array

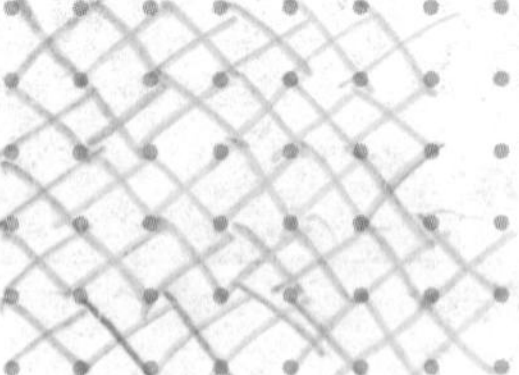

Number sentence: 5x6=30

2. Factors I am using: 4 and 2

First Array

Number sentence: 4x2=8

Second Array

Number sentence: 2x4=8

What do you notice about each pair of arrays? That the product is the same even if they switch.

Math Boxes

Lesson 3-10

DATE TIME

Math Boxes

1 Make equal groups.

14 days make 2 weeks.

Division number model:

14 ÷ 7 = 2

35 days make 5 weeks.

Division number model:

35 ÷ 7 = 5

2 Write each number in expanded form.

498 400+90+8

901 900+00+1

650 600+50+0

762 700+60+2

3 Multiply.

18 = 2 × 9

9 × 1 = 9

9 × 5 = 45

0 = 0 × 9

10 × 9 = 90

4 Solve.

672 + 95 = ?

Unit
numbers

Fill in the circle next to the correct answer.

Ⓐ 623 Ⓑ 667

Ⓒ 727 Ⓓ 767 (filled in)

Think: Does my answer make sense?

5 Use data from the tally chart to finish the picture graph.

Name	Ticket Sales
Rachel	卌 卌 卌
Anna	卌 卌
Chris	卌 卌 卌 卌
Dane	卌

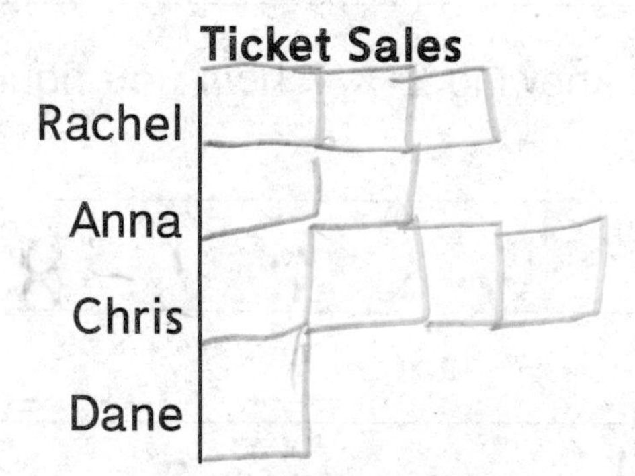

SRB
193-194

Adding a Group

DATE TIME

1 Sketch a picture to show 2 rows of jars with 7 jars in each row. How many jars are there in all?

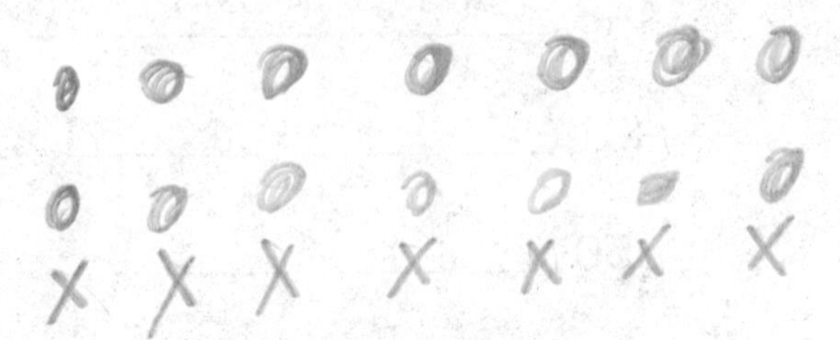

Multiplication number model: 2 × 7 = 14

Use a colored pencil to add another row of jars to show **3 rows** of 7 jars in each row.

I added one group of 7 jars.

Write a number model to describe your new picture.

3 × 7 = 21

How did knowing 2 × 7 help you figure out 3 × 7?

I know that 2x7=14 and if I add a group (add 7) that eq

Adding a Group (continued)

2. Suppose you do not know the answer to $6 \times 3 = ?$.

Helper fact: $5 \times 3 =$ 15

Use pictures, numbers, or words to show how you can use 5×3 to help you figure out 6×3. Solve.

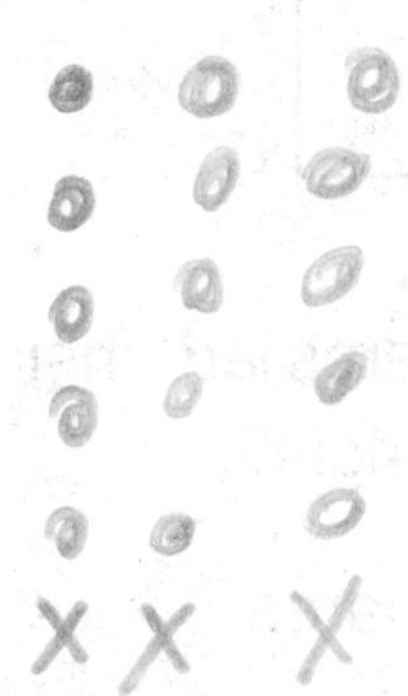

$6 \times 3 =$ 18

Try This

Write a fact you do not know. Then write a helper fact that can help you solve it.

Fact: 6 × 8 = ?

Helper Fact: 5 × 8 = 40

Use pictures, numbers, or words to show how you can use your helper fact to solve a new fact. Solve.

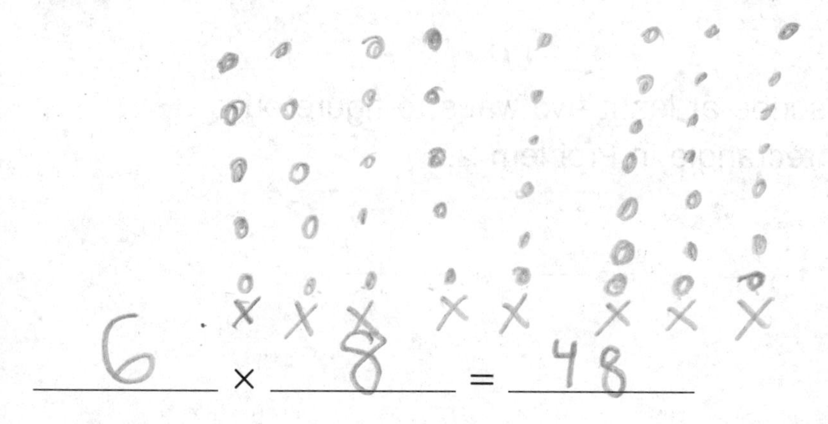

6 × 8 = 48

Math Boxes

1 Measure the length of your thumb to the nearest inch.

about 1 inches

Measure it again to the nearest half inch.

about ______ inches

SRB 171-172

2 Divide the rectangle into 2 rows with 4 same-size squares in each row.

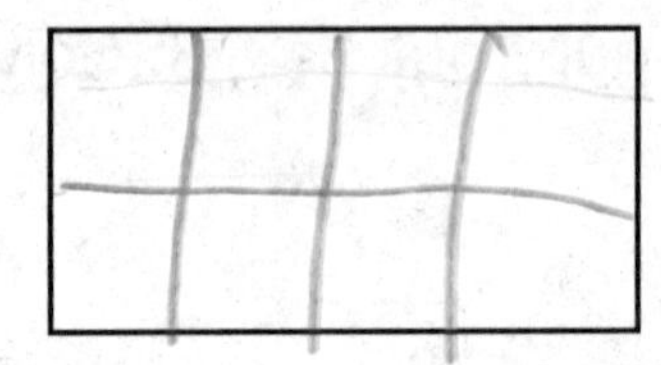

How many squares are inside the larger rectangle?

8 squares

3 Write at least 2 things to describe this shape. Use mathematical language.

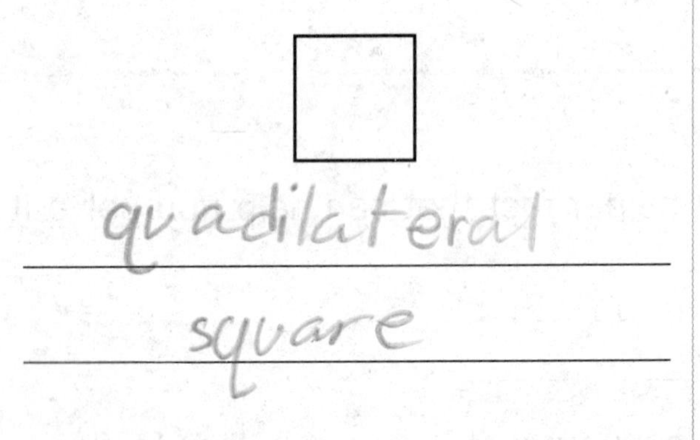

quadilateral

square

______

SRB 216

4 ☐ = 1 square centimeter.

Count the squares.

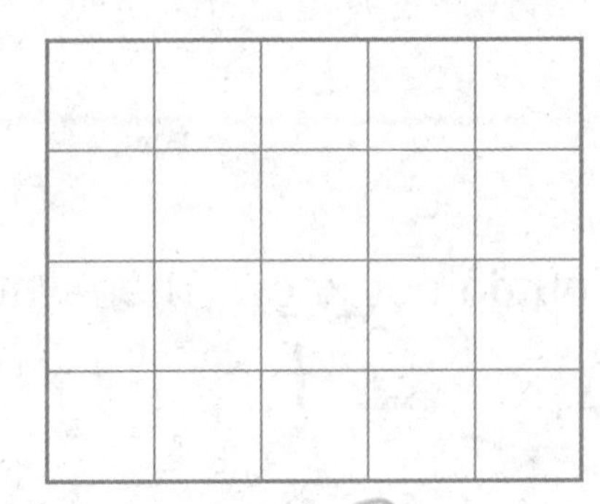

Area: 20 square centimeters

5 **Writing/Reasoning** Describe at least two ways to figure out the number of squares in the rectangle in Problem 2.

$2 \times 4 = 8$ $2+2+2+2=8$

Subtracting a Group

Lesson 3-12

DATE TIME

1 Sketch a picture to show 10 toy cars with 4 wheels on each car. How many wheels are there in all?

  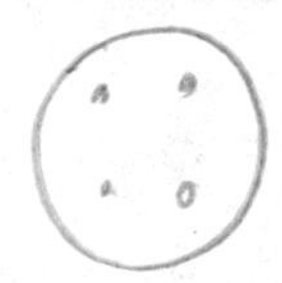 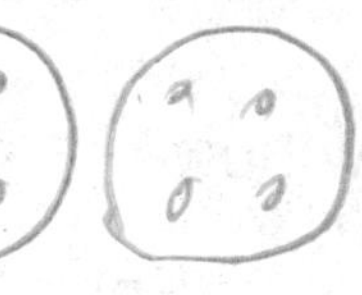 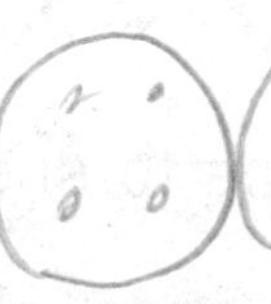  

Multiplication number model: 10 × 4 = 40

Use a colored pencil to change your picture to show the number of wheels on **9 toy cars** with 4 wheels on each car.

Explain what you did.

I subtracted one car from 40.
I know one car equals to 4
so 40 – 4 = 36.

Write a number model to describe your new picture.

9x4=36

How did knowing 10 × 4 help you figure out 9 × 4?

You can just minus 4 from 40.

Subtracting a Group (continued)

2 Suppose you do not know the answer to $4 \times 7 = ?$.

Helper fact: $5 \times 7 =$ 35

Use pictures, numbers, or words to show how you can use 5×7 to help figure out 4×7. Solve.

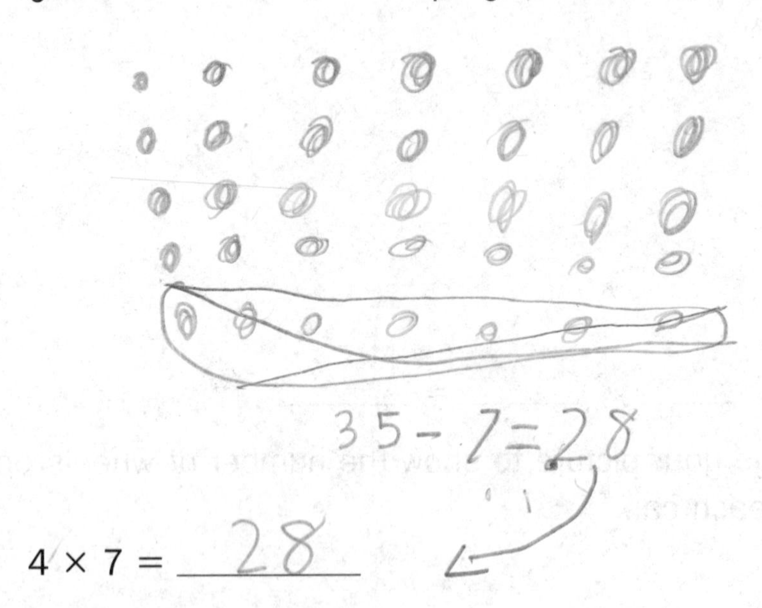

35 - 7 = 28

$4 \times 7 =$ 28

Try This

Write a fact you do not know. Then write a helper fact you can use to help solve it.

Fact: 6 × 8

Helper fact: 5 × 8 = 40

Use pictures, numbers, or words to show how you can use your helper fact to solve an unknown fact. Solve.

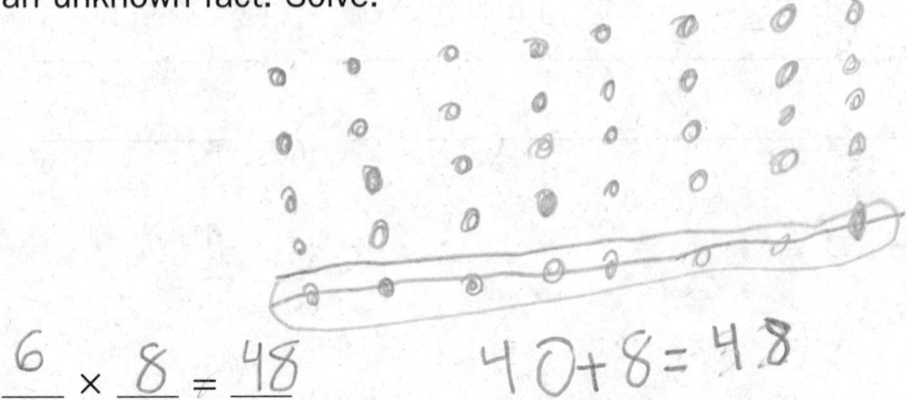

40 + 8 = 48

6 × 8 = 48

Math Boxes

Lesson 3-12

DATE TIME

1. Luis went to his cousin's house at 9:05 A.M. He stayed for 45 minutes. What time did he leave his cousin's house? You may use your toolkit clock or draw an open number line.

Answer: 9:50 am
(unit)

SRB 18-19, 187-188

2. Fill in the missing numbers.

in

Rule
× 10

out

in	out
0	0
1	10
8	80
10	100
500	5,000

SRB 74-75

3. Round to the nearest 100 to make an estimate. Then solve.

Unit
laptops

Estimate:
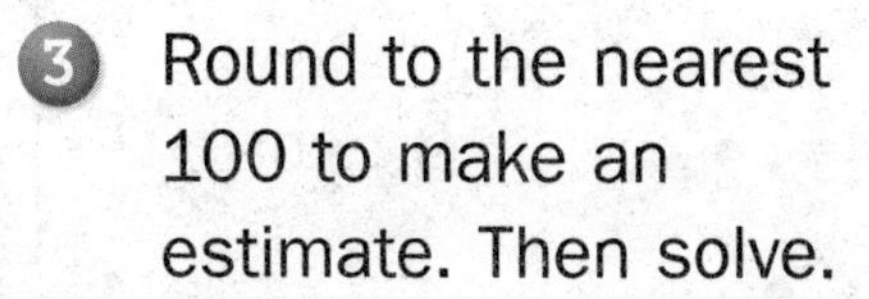

200-100=100

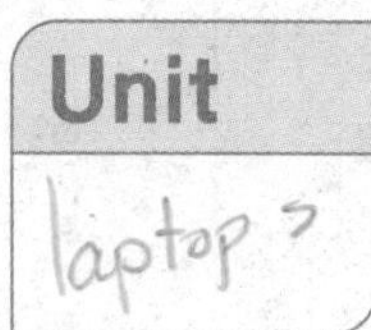

$$\begin{array}{r} 245 \\ -\ 133 \\ \hline 112 \end{array}$$

Think: Does my answer make sense? yes

SRB 106, 119, 122-123

4. Use the dots below to show a 3-by-7 array.

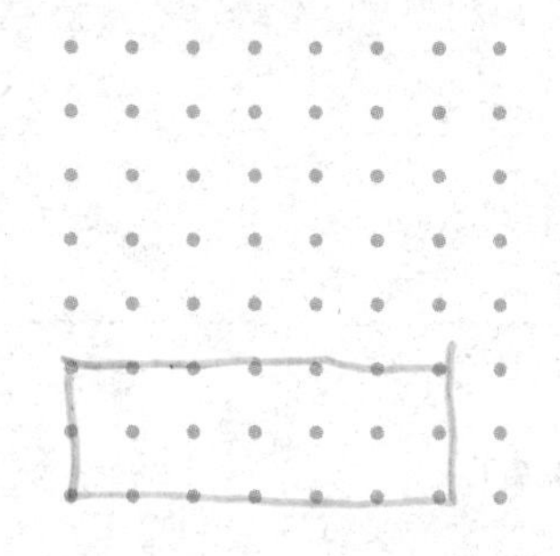

Write a number sentence to match your array.

3×7=21

SRB 41-42

5. **Writing/Reasoning** Explain how you can use your estimate to check your answer for Problem 3.

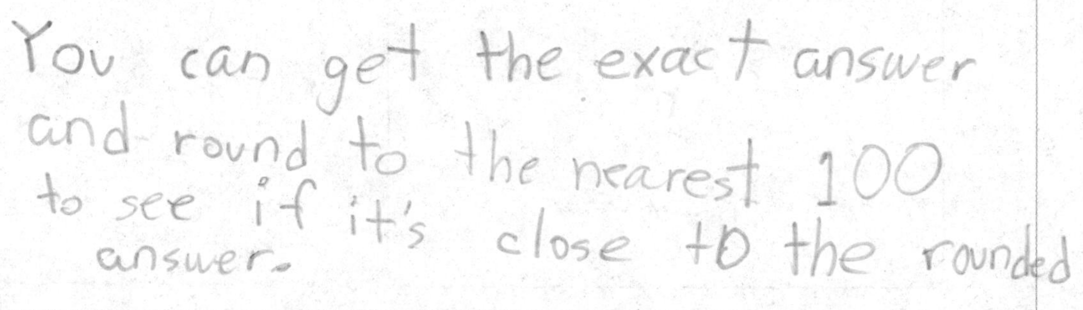

You can get the exact answer and round to the nearest 100 to see if it's close to the rounded answer.

Name-Collection Boxes

Lesson 3-13

DATE TIME

Work with a partner. Use addition, subtraction, multiplication, and division to make your number names.

1. Write at least 10 names in the box.

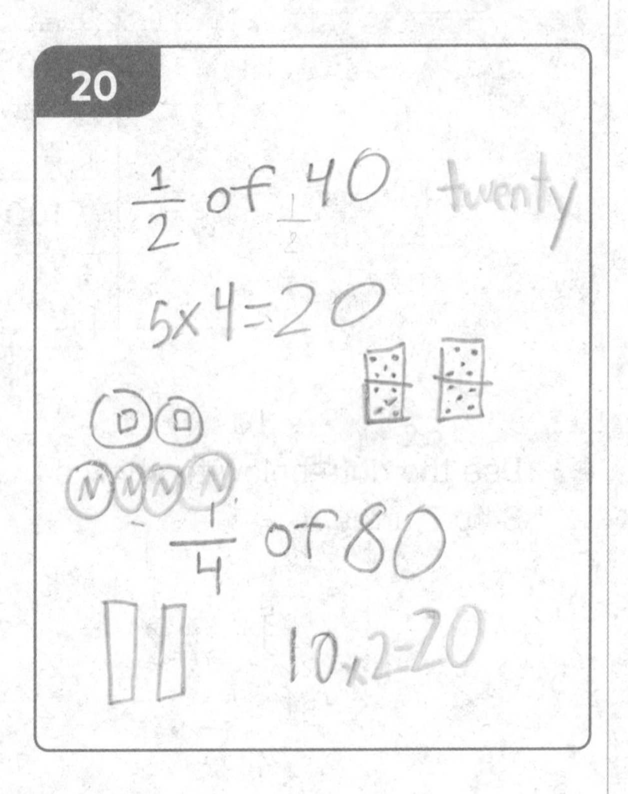

2. Three names do not belong in this box. Cross them out. Then write the name of the box on the tag.

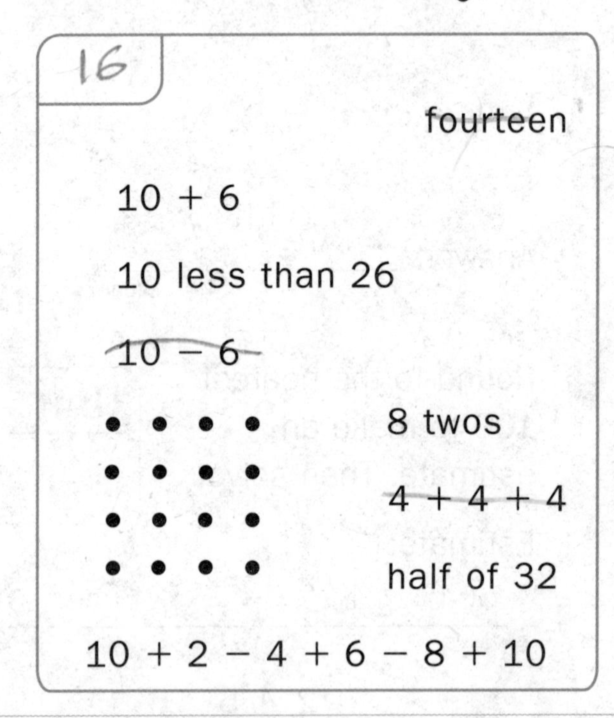

16
fourteen
10 + 6
10 less than 26
10 − 6
8 twos
4 + 4 + 4
half of 32
10 + 2 − 4 + 6 − 8 + 10

Do this on your own.

3. Write at least 10 names in the box.

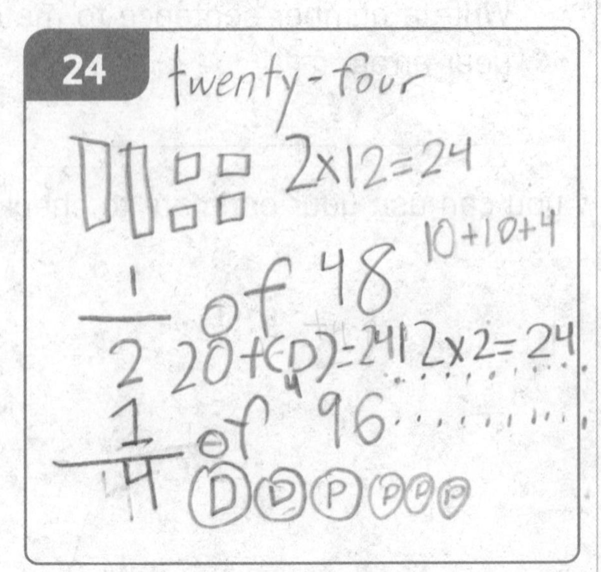

Do this on your own.

4. Make up your own box. Write at least 10 names.

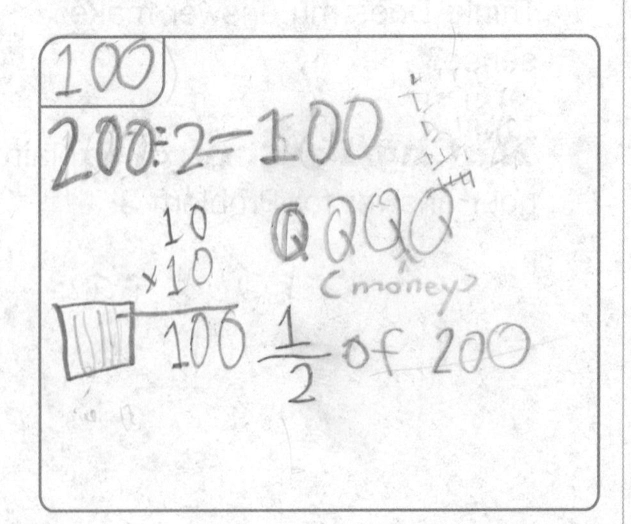

Measuring Line Segments

Lesson 4-1

DATE TIME

Use Ruler A to measure each line segment to the nearest inch (in.).

Use Ruler B to measure each line segment to the nearest centimeter (cm).

	Ruler A	Ruler B
1 ________	about 3 in.	about 8 cm
2 ________	about 5 in.	about 13 cm

Use Ruler A to measure each line segment to the nearest $\frac{1}{2}$ inch (in.).

Use Ruler B to measure each line segment to the nearest centimeter (cm).

	Ruler A	Ruler B
3 ________	about $3\frac{1}{2}$ in.	about 9 cm
4 ________	about $4\frac{1}{2}$ in.	about 11 cm
5 ________	about 4 in.	about 10 cm

6 Explain how you used Ruler A to measure to the nearest $\frac{1}{2}$ inch in Problem 3.

I saw the line ended half way between the 3 and the 4 on the inch ruler.

...ual Rulers

Lesson 4-1

DATE TIME

...e rulers. Think about which rulers will help you make correct measurements. Then answer the questions.

Ruler 1

Ruler 2

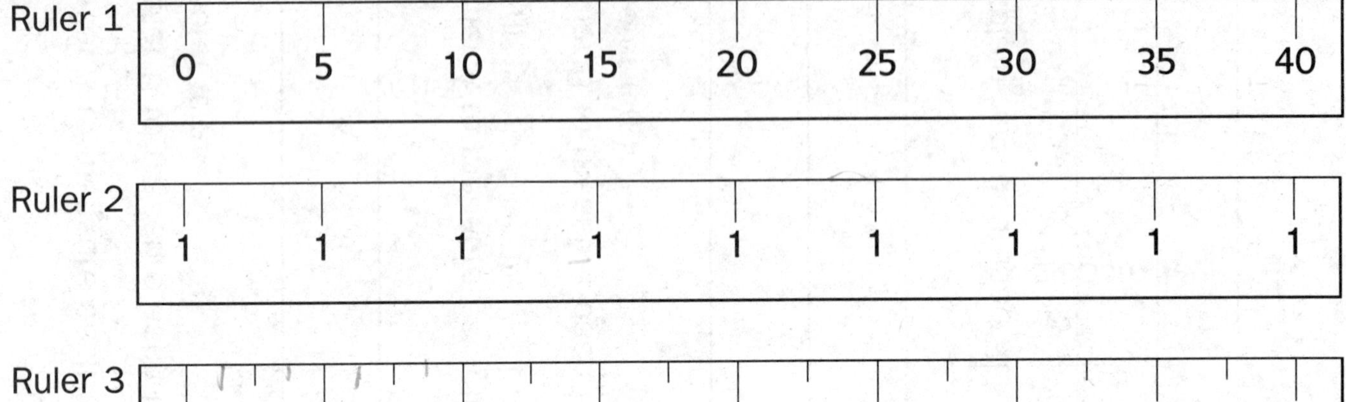

Ruler 3

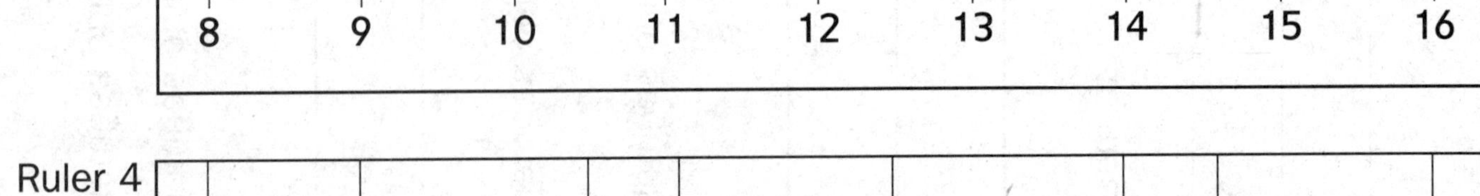

Ruler 4

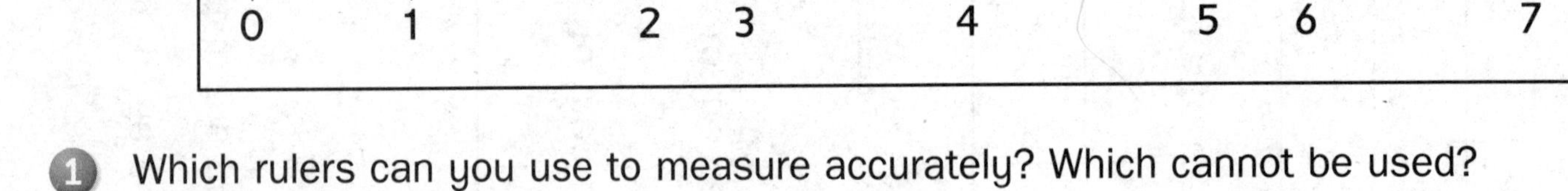

1. Which rulers can you use to measure accurately? Which cannot be used? Explain your answers.

____

____

____

2. How could you use Ruler 2 to measure the length of a pencil?

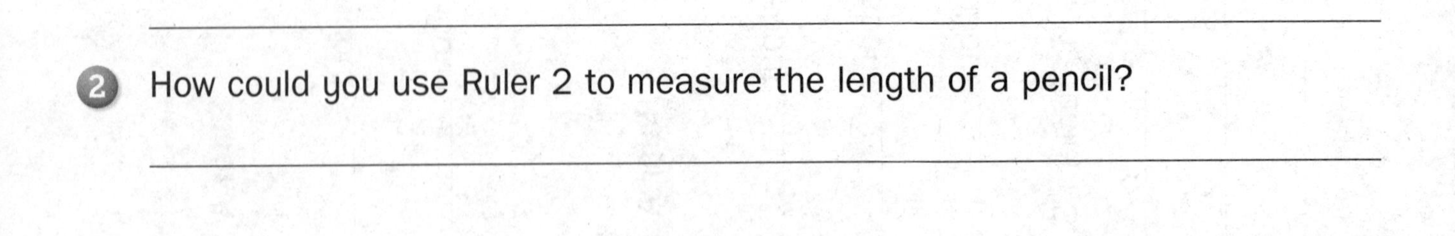

3. Which of the above rulers could be used to make more precise measurements?

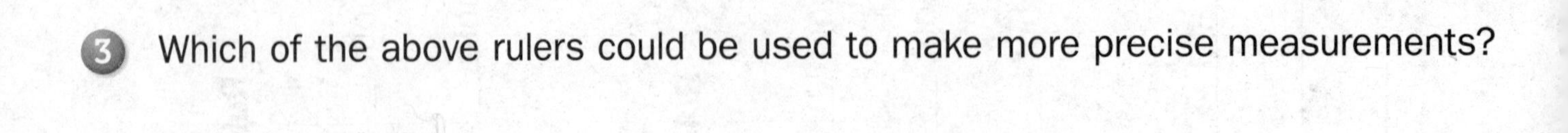

Number Stories with Length

Lesson 4-1

DATE TIME

Read each number story. Think about the steps you need to take to solve. You may draw pictures to help. Write number models to show your thinking.

1. Mr. Miller is building a dog pen and needs 42 feet of fence. He has 2 pieces of fence that are each 12 feet long. How many more feet of fence does he need?

Number models: $12 \times 2 = 24$ $42 - 24 = 18$

Answer: 18 feet (unit)

2. Trey runs 3 miles, 5 days a week. How many miles does Trey run in 2 weeks?

Number models: $3 \times 5 = 15$ $15 + 15 = 30$

Answer: 30 miles (unit)

Try This

3. It takes 3 feet of fabric to make one book bag and 2 feet of fabric to make the straps for one book bag. Lola has 25 feet of fabric. How many book bags with straps can she make?

Number models: $3 + 2 = 5$ $25 \div 5 = 5$

Answer: 5 bags (unit)

Math Boxes

DATE TIME

Math Boxes

1. The third-grade class starts lunch at 10:55 A.M. They have 40 minutes to eat. What time is lunch over?

11:35 A.M.

SRB 18-19, 187-188

2. Fill in the missing numbers.

in

Rule

÷ 5

out

in	out
5	1
10	2
25	5
30	6
45	9
100	20

SRB 74-75

3. An alligator clutch had 82 eggs. 19 eggs did not hatch. How many eggs did hatch?

82 − 19 = ?

(number model with ?)

Answer: 63 eggs (unit)

SRB 30-31

4. Use this array to show how 5 × 3 can help you figure out 6 × 3.

× × ×
× × ×
× × ×
× × ×
× × ×
× × ×

Helper fact: 5 × 3 = 15

6 × 3 = 18

SRB 47

5. **Writing/Reasoning** What strategy could you use to check your answer to Problem 3?

The counting-up strategy.

Shoe-Length Data

Lesson 4-2

DATE TIME

Look at the measurements on the stick-on notes.

1. What is the shortest shoe length in your class? 8"
2. What is the longest shoe length in your class? 10"

Use the class shoe-length data to complete the line plot.

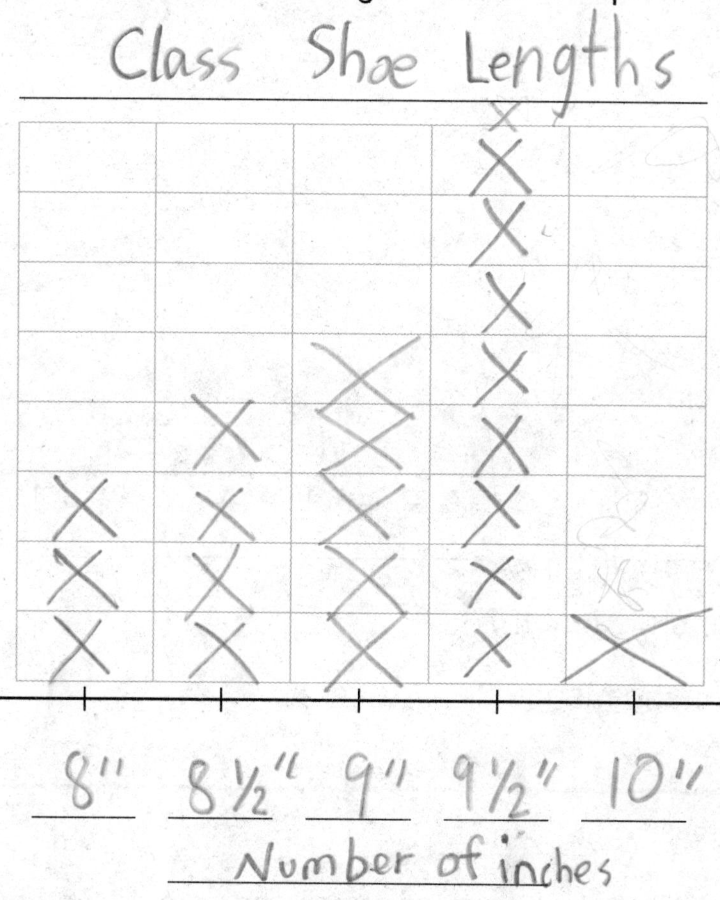

3. If you were buying gym shoes for your class, which sizes should you buy the most of? Why?

I would buy 9 1/2 in. shoes because that's the most popular.

Fourth-Grade Shoe Lengths

Lesson 4-2

DATE TIME

A fourth-grade class is going to use their shoe-length data to buy gym shoes for the entire fourth grade. Use their line plot to answer the questions.

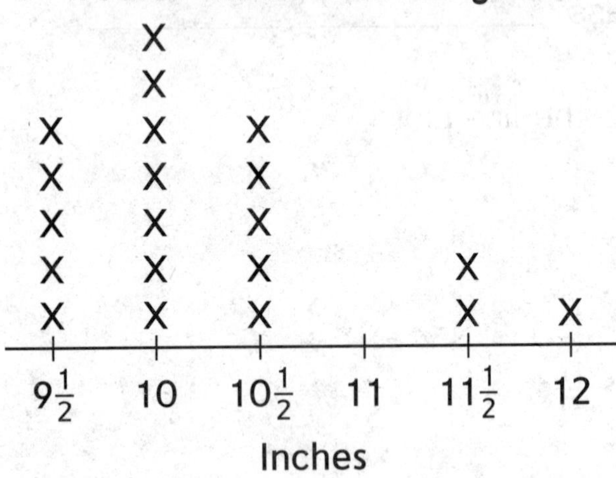

1. How many children are in this class? 20

   How do you know? If you count all of the Xs, you will find how much children is there.

2. What is the longest (maximum) shoe length? 12

3. What is the shortest (minimum) shoe length? 9 1/2

Try This

4. Aubrey thinks the class should buy a few pairs of shoes that are 11 inches long, although no one in the class has that shoe length. Do you agree? Explain.

   Yes because no one has that shoe size.

Creating a Picture Graph

Lesson 4-2

DATE TIME

The chart to the right shows the approximate average annual snowfall, in inches, for five of the largest U.S. cities that receive at least 5 inches of snow per year.

Draw a picture symbol on the line next to "KEY" to show 5 inches of snow. Complete the picture graph using the key and the data in the chart.

City	Average Annual Snowfall
New York	30 inches
Chicago	40 inches
Philadelphia	20 inches
Detroit	40 inches
Indianapolis	25 inches

Write a title for your picture graph.

Title: __________

New York	X	X	X	X	X	X			
Chicago	X	X	X	X	X	X	X	X	
Philadelphia	X	X	X	X					
Detroit	X	X	X	X	X	X	X	X	
Indianapolis	X	X	X	X	X				

KEY: X = 5 inches of snow

1. Look at the picture graph. How much more does it snow, on average, in Chicago than in Indianapolis? 15 inches

2. Write a different question that can be answered from the picture graph.
How many more inches are thee in New York and Philidelphia?

Math Boxes

Lesson 4-2

DATE TIME

Math Boxes

1. Santiago's mom doubles his $10 allowance after he does extra chores. He spends $5. How much money does he have left?

Number models: 2 x 10 = 20 20 - 5 = 15

Answer: 15 dollars (unit)

SRB 30-31

2.

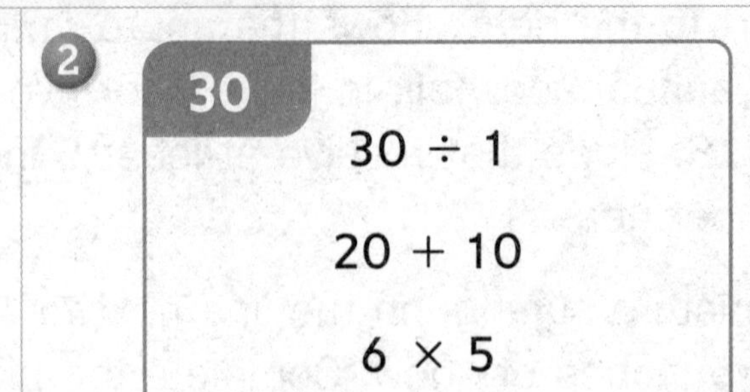

Which other names can go in the name-collection box? Circle all that apply.

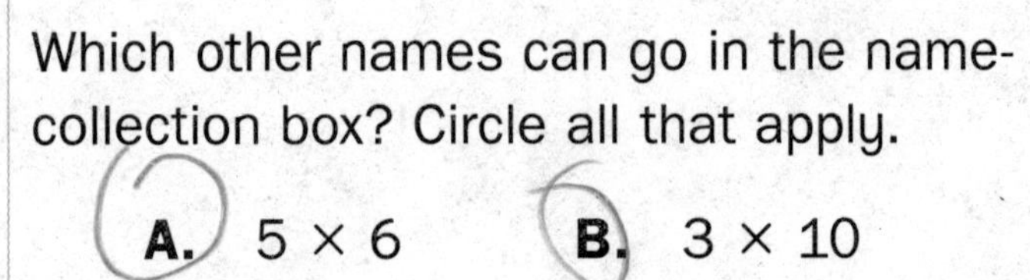

SRB 96-97

3. Use the 5-by-6 array to show how you can use 5 × 6 to help you figure out 4 × 6.

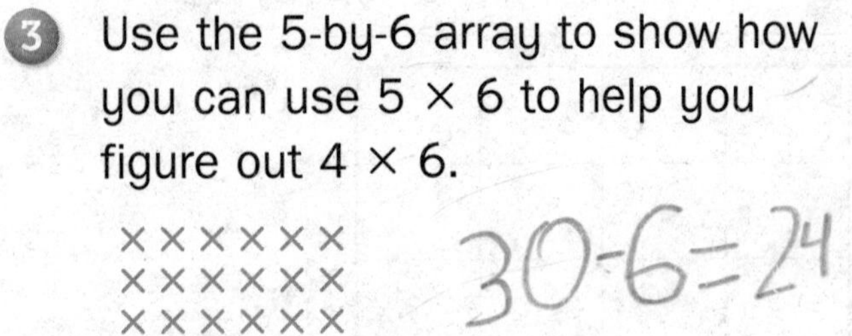

Helper fact: 5 × 6 = 30

4 × 6 = 24

SRB 48

4. Round to the nearest 10 and make an estimate. Then solve.

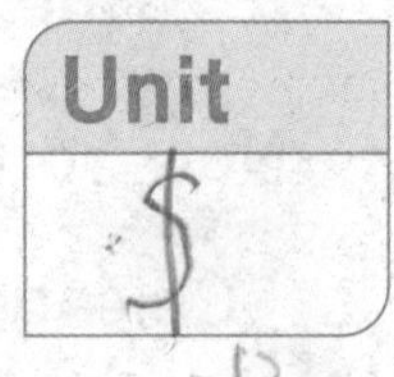

Estimate:

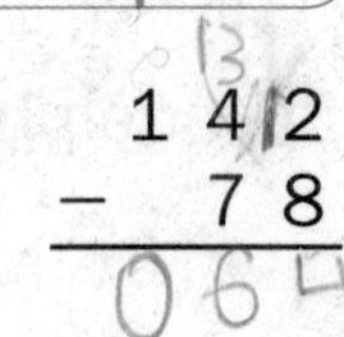

Think: Does my answer make sense?

5. Use the data in the tally chart to finish the picture graph.

Favorite Pizza Topping	Number of Votes
Mushrooms	////
Cheese	~~////~~ ~~////~~
Pepperoni	~~////~~ //
Green Peppers	///

How many more children voted for cheese than for mushrooms and green peppers combined? 3

SRB 193-194

Measuring Distances Around Objects

Lesson 4-3

DATE TIME

Measure the distances around some small objects and some large objects to the nearest $\frac{1}{2}$ inch.

1. Object: eraser Measurement: about $2\frac{1}{2}$ inches
2. Object: SRB Measurement: about 19 inches
3. Object: dry erase marker Measurement: about 2 inches
4. Object: post its Measurement: about $7\frac{1}{2}$ inches
5. Object: book Measurement: about $11\frac{1}{2}$ inches
6. Object: tape measure Measurement: about 6 inches
7. How would you measure the distance around a real sign that looks like the one in the picture below?

By measuring a place wher

Comparing Masses

Lesson 4-3

DATE TIME

Use the pan balance and set of standard masses to find objects that have the following masses:

about 1 gram centimeter cube

about 50 grams pack of posted notes

about 100 grams nine markers

about 500 grams ______

about 1,000 grams ______

Find another object in your classroom that has about the same mass as one of the standard masses.

Object: ______ Mass: about ______ grams

Choose one of the objects listed above. Explain how you can use that object to estimate the mass of another object.

______

______

______

______

______

Math Boxes

1 The third graders have recess for 25 minutes. If recess begins at 12:40 P.M., what time is recess over?

1:05 p.m.

40 + 25 = 65

SRB 18-19, 187-188

2 Fill in the missing numbers.

in → Rule: −9 → out

in	out
10	1
20	11
13	4
14	5
100	91
59	50

SRB 74-75

3 Luca had $261 in his bank. He took some money out to buy a bike. He had $109 left. How much did he pay for his bike? You may draw a diagram.

$261 − 109 = ?

(number model with ?)

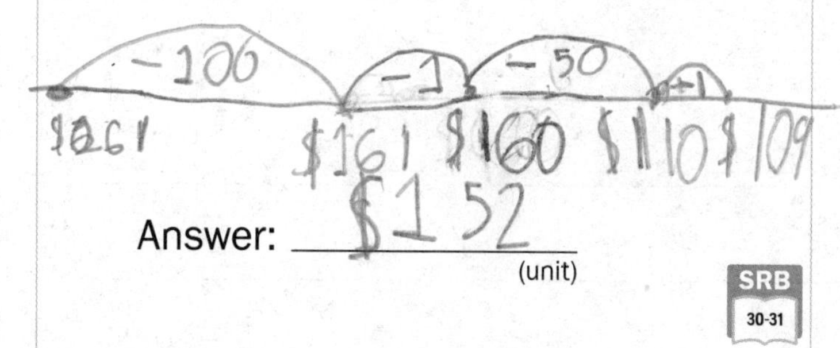

Answer: $152

(unit)

SRB 30-31

4 Use this array to show how 5 × 4 can help you figure out 6 × 4.

× × × ×
× × × ×
× × × ×
× × × ×
× × × ×
X X X X

20 + 4 = 24

Helper fact: 5 × 4 = 20

6 × 4 = 24

5 **Writing/Reasoning** In Problem 4, how might the helper fact, 5 × 4 = 20, help you figure out 4 × 4?

You can minus a row (−4) like 20 − 4.

Which Does Not Belong?

Lesson 4-4

DATE TIME

Math Message

1. Look at the four shapes below. Circle the shape that does not belong.

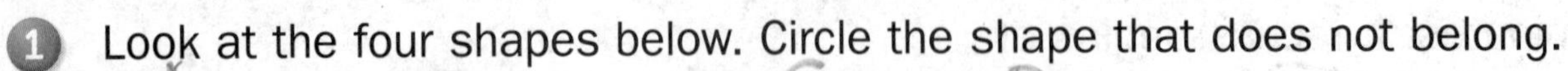

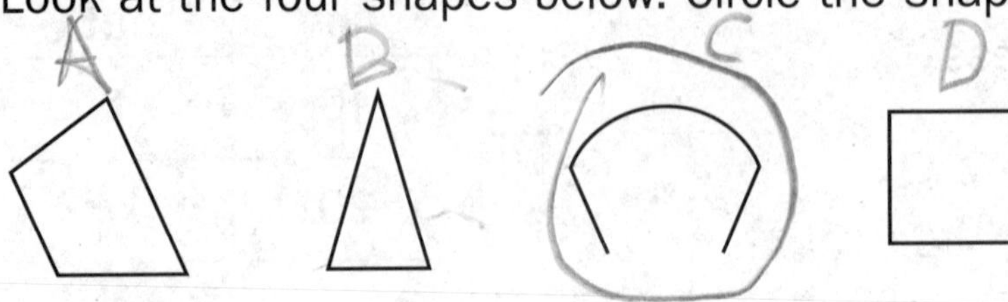

Explain your answer.

C is not a polygon because it is not closed.

2. Look at the four shapes below. Circle the shape that does not belong.

Explain your answer.

B is not a polygon because it is not closed.

Math Boxes

Lesson 4-4

DATE TIME

1. Alex had 6 packs of pencils with 6 pencils in each pack. He gave 1 pack away. Have many pencils does he have now? Write number models to help keep track of your thinking.

Number models: $6 \times 6 = 36$ $36 - 6 = 30$

Answer: 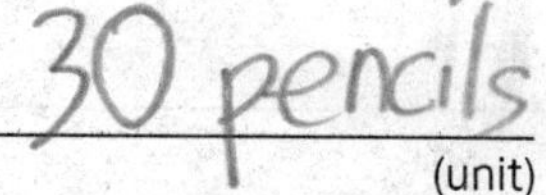 (unit)

SRB 30-31

2. Three names do not belong. Cross them out. Then write the name of the box on the tag.

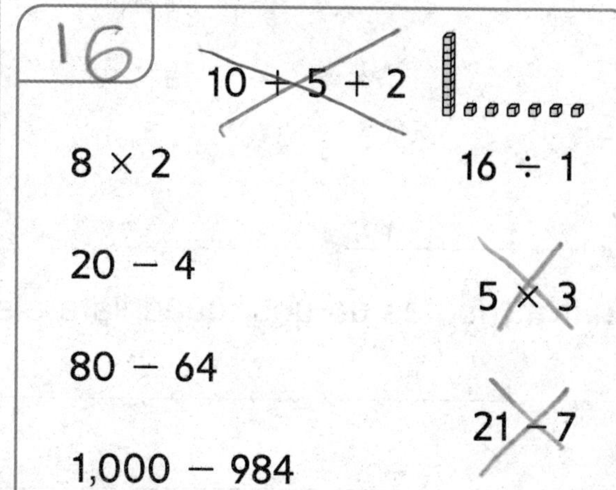

SRB 96-97

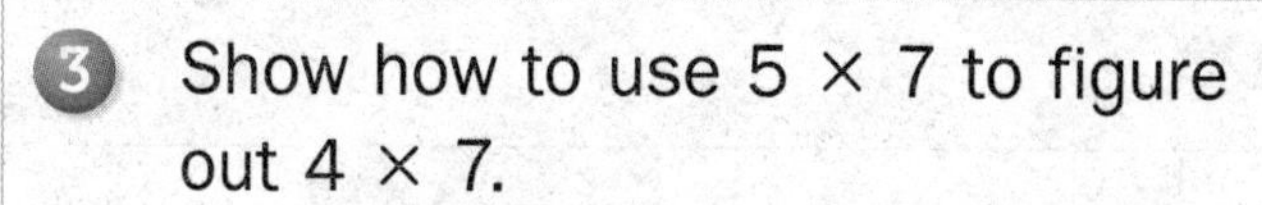

3. Show how to use 5×7 to figure out 4×7.

× × × × × × ×
× × × × × × ×
× × × × × × ×
× × × × × × ×
× × × × × × ×

Helper fact:
$5 \times 7 = 35$

$35 - 7 = 28$

$4 \times 7 =$ 28

SRB 48

4. Subtract.
$332 - 159 = ?$

- ● 173
- ○ 183
- ○ 273
- ○ 227

Think: Does my answer make sense? $300 - 200 = 100$

SRB 119, 122-123

5. Use the data in the tally chart to finish the picture graph.

Day of Week	Number of Books
Monday	𝍸 𝍸 𝍸 𝍸 𝍸 𝍸
Tuesday	𝍸 𝍸 𝍸 𝍸 𝍸
Wednesday	𝍸 𝍸 𝍸 𝍸
Thursday	𝍸 𝍸 𝍸
Friday	𝍸 𝍸

Number of Books Checked Out

Monday
Tuesday
Wednesday
Thursday
Friday

Key: □ = 10 books

How many more books were checked out on Wednesday and Thursday together than on Friday?

25 books (unit)

SRB 193-194

Quadrilateral Relationships

DATE TIME

1. Sketch your two quadrilaterals below. Record the type of quadrilaterals they are on the lines below.

What attributes do your quadrilaterals have in common?

What attributes are different?

2. Sketch two more of your quadrilaterals below. Record the type of quadrilaterals they are on the lines below.

What attributes do your quadrilaterals have in common?

What attributes are different?

Thaman

DATE 1/10/20 TIME

1 Round to the nearest 100 and make an estimate. Then solve. Show your work.

Unit
toys

Estimate: 600 + 100 = 700

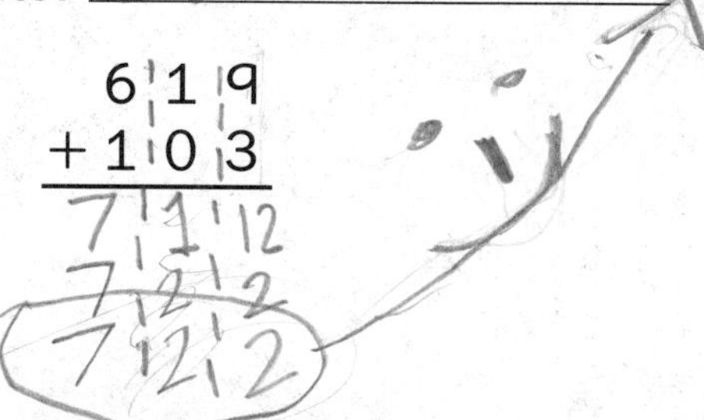

6 1 9
+ 1 0 3
7 1 12
7 2 2

Think: Does my answer make sense? Yes

2 Draw an array that has 6 rows and 6 Xs in each row. Write a number sentence for your array.

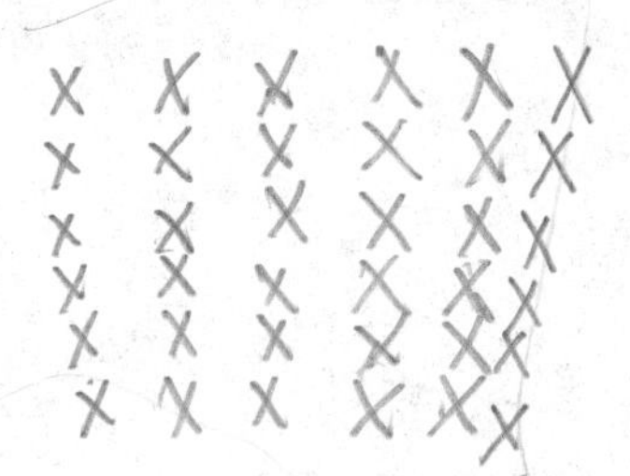

Number sentence: 6 × 6 = 36

3 Draw a 2-by-5 array. Then draw a 5-by-2 array.

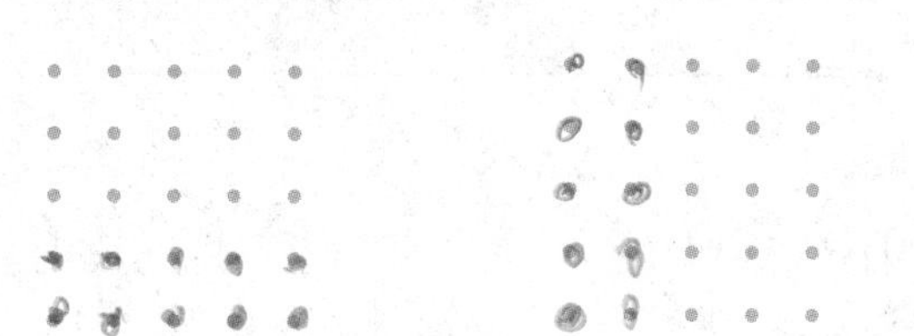

Complete the number sentences.

2 × 5 = 10   5 × 2 = 10

4 Measure the length of this line segment to the nearest centimeter. Choose the correct answer.

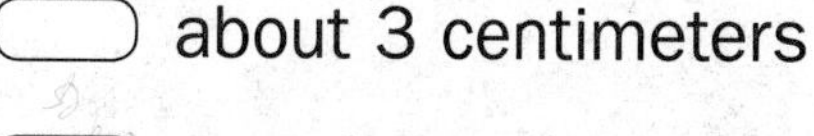
about 3 centimeters

about 6 centimeters

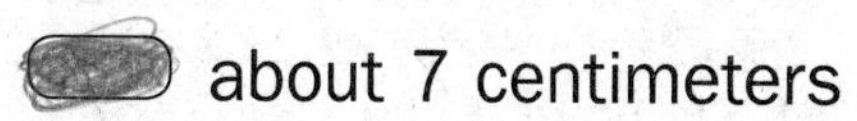
about 7 centimeters

about 10 centimeters

5 **Writing/Reasoning** How are your arrays similar in Problem 3? How are they different?

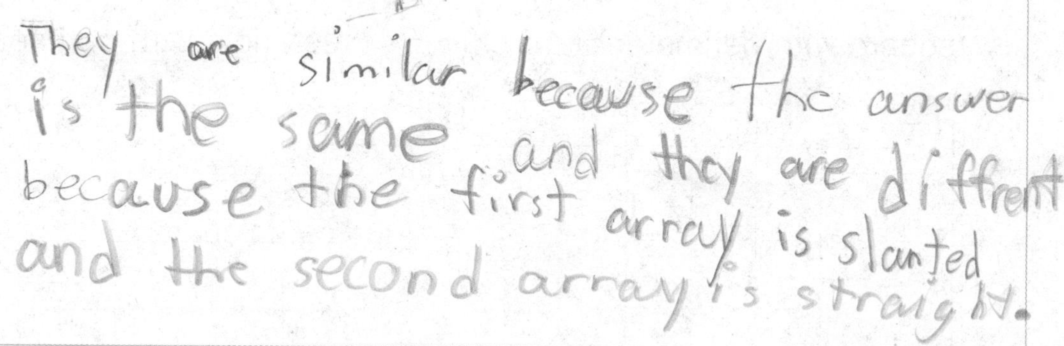

They are similar because the answer is the same and they are different because the first array is slanted and the second array is straight.

Measuring Perimeters of Polygons

Lesson 4-6

DATE TIME

Measure the sides of each polygon to the nearest half inch.

Use the side lengths to find the perimeters.

Write a number sentence to show how you found the perimeter.

1

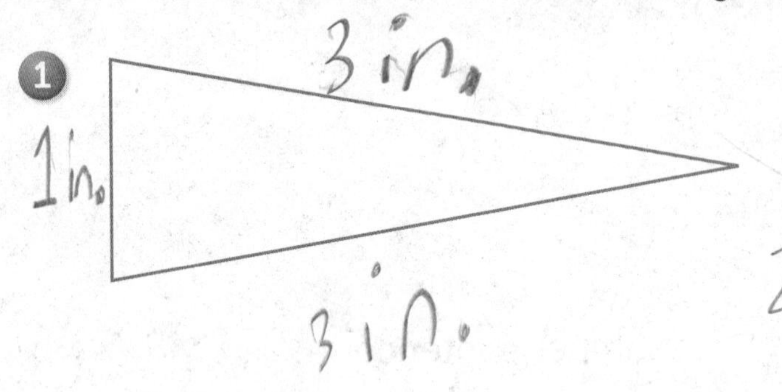

Number sentence: 3+3+1=7

Perimeter: about 7 inches

2

3 in.

2 in.

3 in.

Number sentence: 3+3+2=8

Perimeter: about 8 inches

3

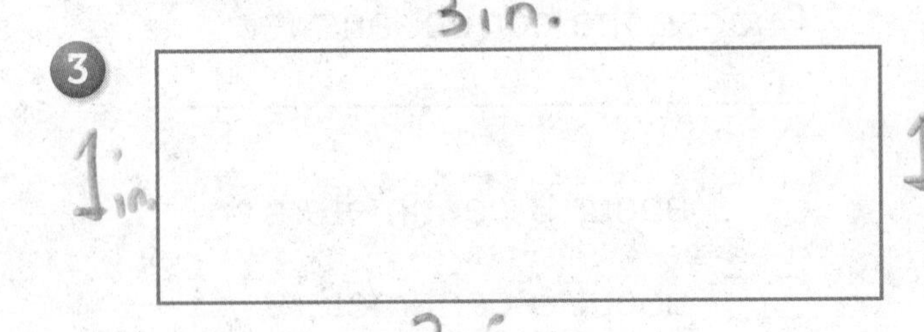

Number sentence: 3+3+1+1=8

Perimeter: about 8 inches

4

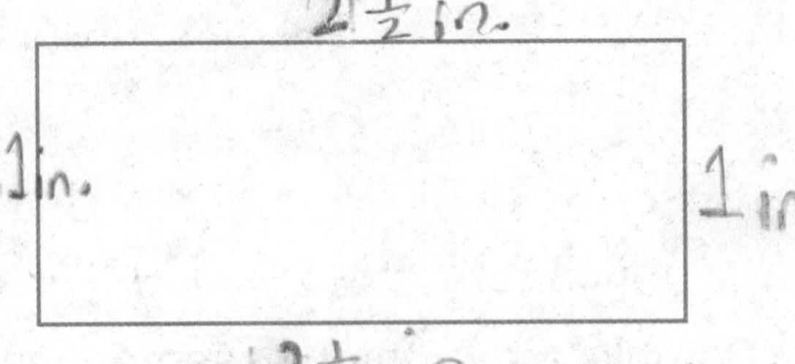

Number sentence: $2\frac{1}{2}+2\frac{1}{2}+1+1=7$

Perimeter: about 7 inches

Try This

5 Draw each shape on the centimeter grid.

square with perimeter = 16 cm

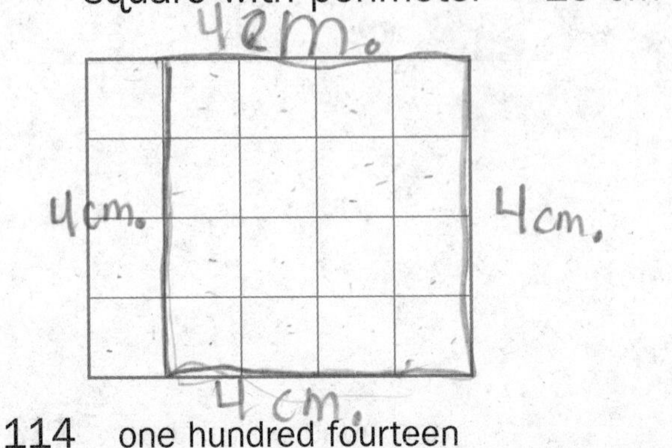

rectangle with perimeter = 20 cm

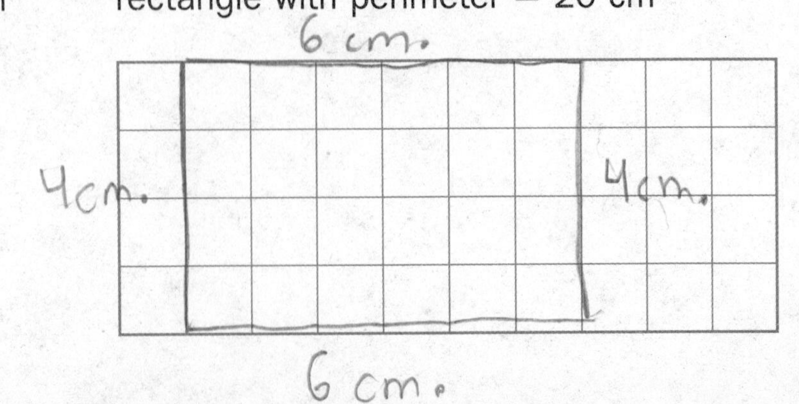

Perimeter Number Stories

Lesson 4-6

DATE TIME

Solve each number story. Show your work.

1. Mrs. McMaster wants to add a border to a rectangular bulletin board. The top is 35 inches across, and the side is 25 inches tall. How much border does Mrs. McMaster need? You may sketch a picture.

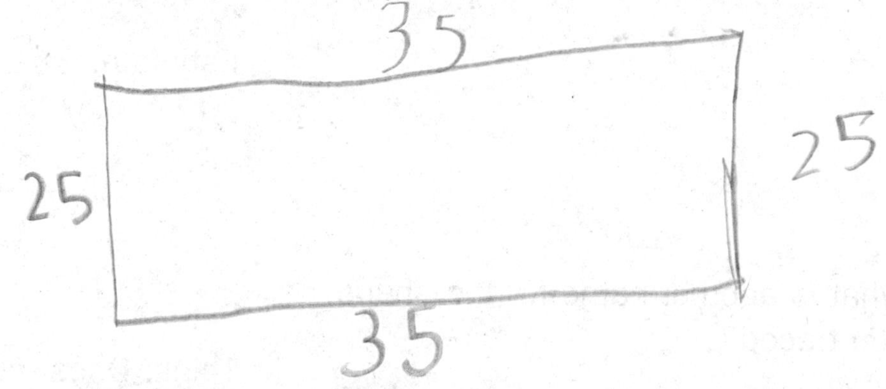

Number model: 35+ 35 + 25 + 25 = 120 in.

Mrs. McMaster needs 120 inches of border.

2. Mr. Lopez wants to put a fence around his rectangular vegetable garden. The longer sides are 14 feet long and the shorter sides are $9\frac{1}{2}$ feet long. How much fencing should Mr. Lopez buy? You may sketch a picture.

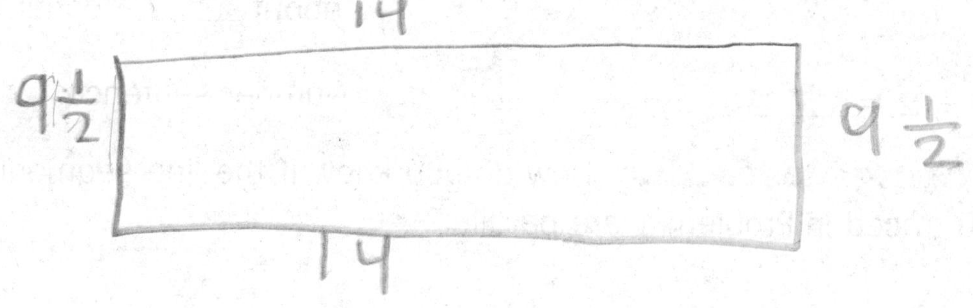

Number model: $14 + 14 + 9\frac{1}{2} + 9\frac{1}{2} = 46$

Mr. Lopez should buy 46 feet of fencing.

Math Boxes

DATE TIME

1 Use your Pattern-Block Template. Trace a parallelogram.

What is another name for the shape you traced?

SRB 216-217

2 Round to the nearest 10 and make an estimate. Then solve. Show your work.

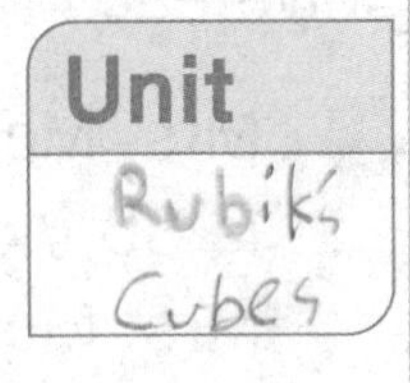

Estimate: 310 − 210 = 100

  3 0 7
− 2 0 9
  1 9 8

Think: Does my answer make sense? almost

3 Draw a sketch to show 18 ÷ 2.

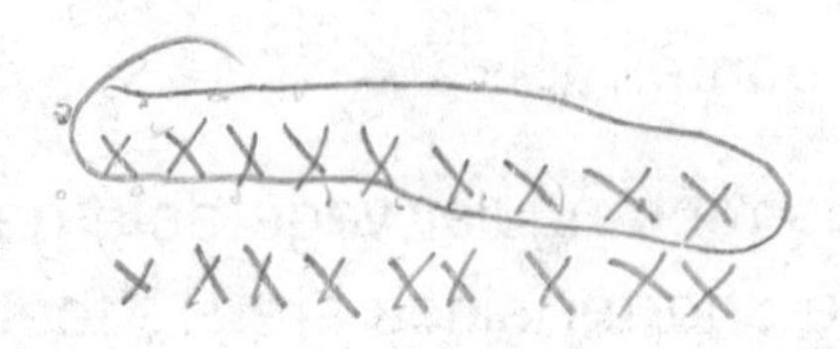

18 ÷ 2 = 9

4 Measure the line segment to the nearest $\frac{1}{2}$ inch and to the nearest centimeter.

about 3 inches

about 7 centimeters

5 **Writing/Reasoning** How do you know if the line segments in the shape you traced in Problem 1 are parallel?

It is parallel because all of the sides are never going to meet any point

SRB 209

Comparing Perimeter and Area

Lesson 4-7

DATE TIME

For Problems 1–3, find the perimeter and the area of the rectangle.

1. 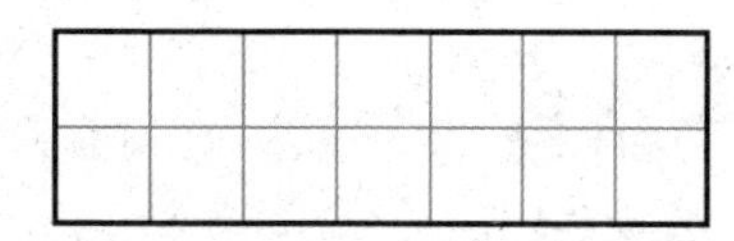

Key: □ = 1 square foot

Perimeter: 18 feet

Area: 14 square feet

2. 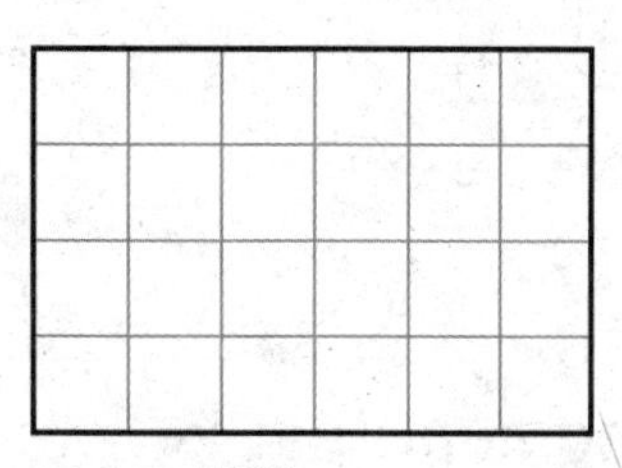

Key: □ = 1 square meter

Perimeter: 20 meters

Area: 24 square meters

3. 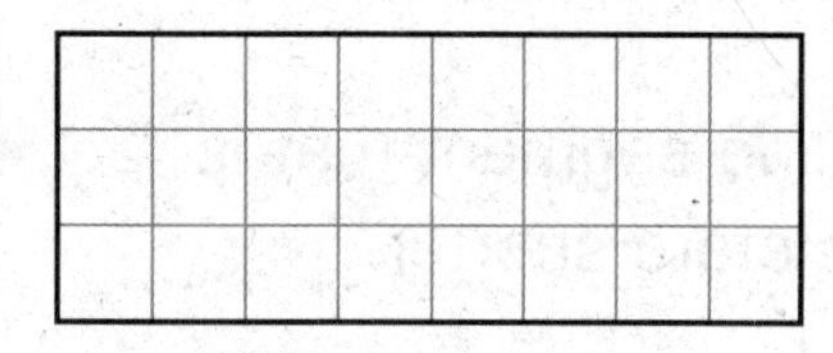

Key: □ = 1 square mile

Perimeter: 22 miles

Area: 24 square miles

Try This

4. Find the perimeter and the area of this shape.

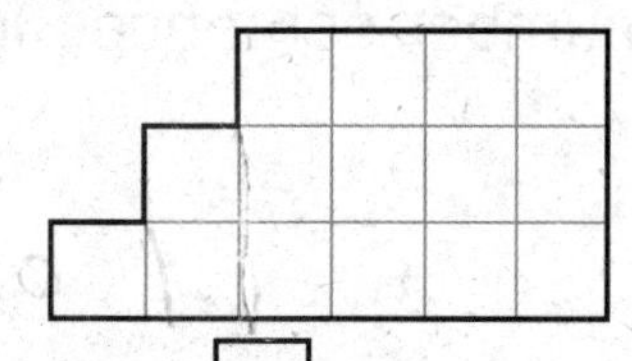

Key: □ = 1 square centimeter

Perimeter: 17 centimeters

Area: 15 square centimeters

5. Nicolas says he can measure both the perimeter and the area of a rectangle using a square. Do you agree or disagree? Explain your answer using words or drawings.

Yes 2. and then 1.

Math Boxes

Lesson 4-7

DATE TIME

1 Solve.

Unit

$328 + 294 = ?$

- ◯ 512
- ◯ 612
- ◯ 622
- ◯ 51,112

SRB 116-118

2 Solve.

$9 =$ ______ $\times 3$

______ $= 4 \times 4$

$7 \times 7 =$ ______

$9 \times$ ______ $= 81$

SRB 44

3 Draw a 3-by-5 array. Then draw a 5-by-3 array. Write a number sentence to match each array.

______ ______

4 Draw a line segment that is 7 centimeters long.

Draw a line segment that is 2 centimeters shorter.

5 **Writing/Reasoning** Draw an array for one of the number sentences in Problem 2. What shape is it and why?

SRB 41, 44

Areas of Rectangles

Lesson 4-8

DATE TIME

Math Message

On the centimeter grid below, draw one rectangle with short sides 3 centimeters and long sides 5 centimeters. Then draw a second rectangle with short sides 1 centimeter and long sides 15 centimeters.

Label the side lengths of both rectangles. Figure out and record the area and perimeter for each rectangle.

☐ = 1 square centimeter

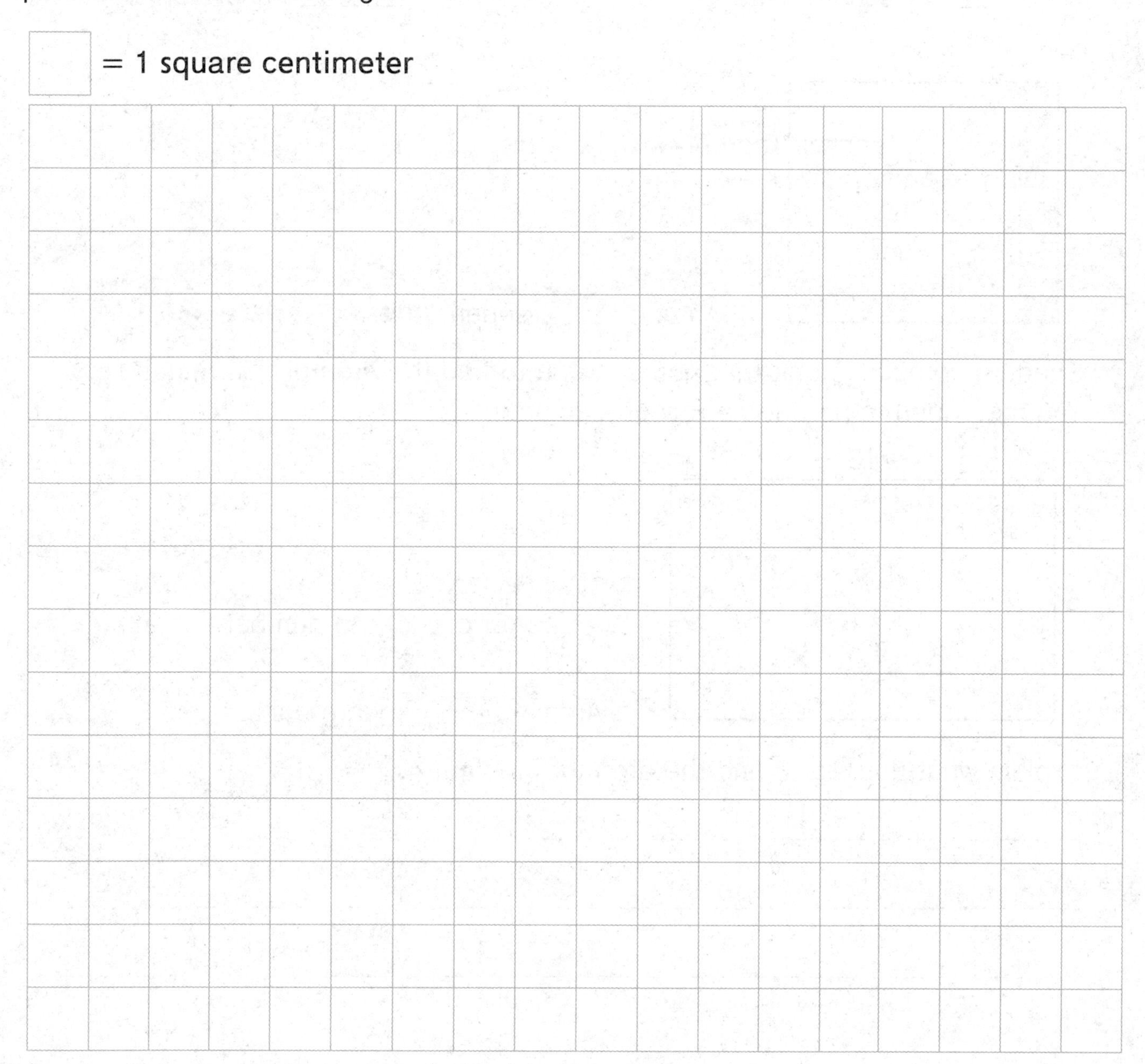

Areas of Rectangles (continued)

Lesson 4-8

DATE TIME

Use the shaded composite unit to find the area of each rectangle.

1. 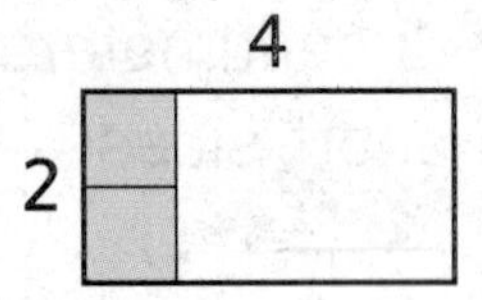

Area: _______ square units

2. 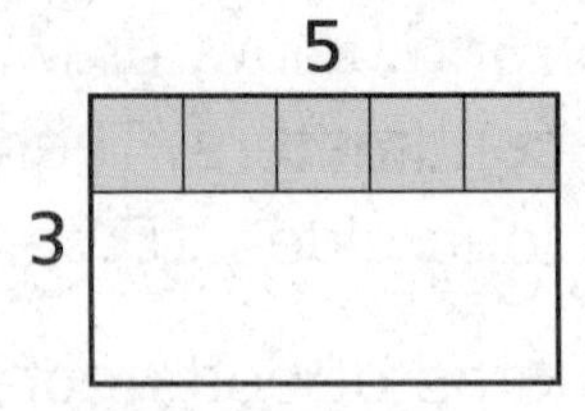

Area: _______ square units

3. 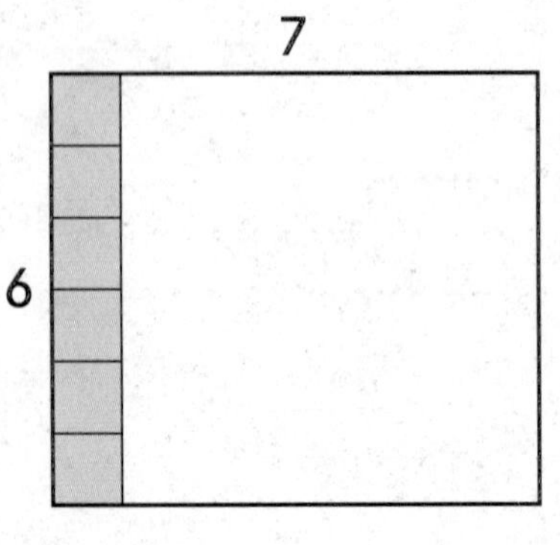

Area: _______ square units

4. Shade a composite unit that you can use to find the area of this rectangle. You may need to partition a row or a column.

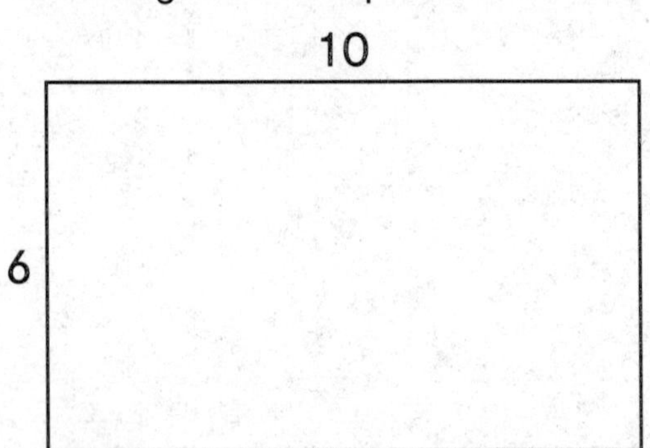

Area: _______ square units

5. Explain what you did to find the area for Problem 4.

____

____

____

Body Measures

Lesson 4-8

DATE TIME

Work with a partner to find each measurement to the nearest $\frac{1}{2}$ inch.

	Me	Partner
Date	____________	____________
height	about ________ in.	about ________ in.
knee to foot	about ________ in.	about ________ in.
around neck	about ________ in.	about ________ in.
waist to floor	about ________ in.	about ________ in.
forearm	about ________ in.	about ________ in.
hand span	about ________ in.	about ________ in.
arm span	about ________ in.	about ________ in.

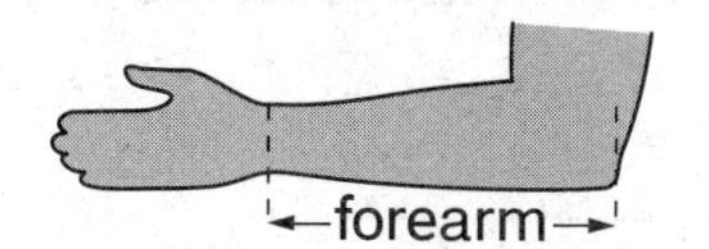

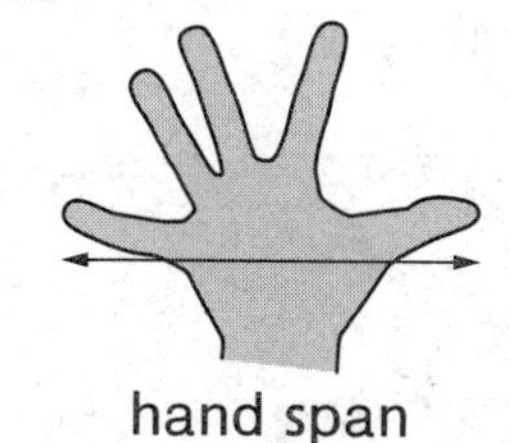

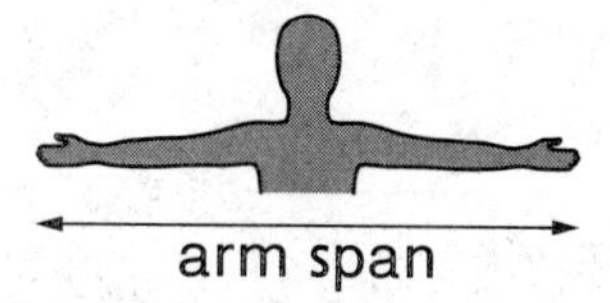

How do you know whether your body measurements make sense?

____

____

____

Math Boxes

1. Draw a quadrilateral with 4 sides that are equal in length.

Another name for your quadrilateral is quadrangle.

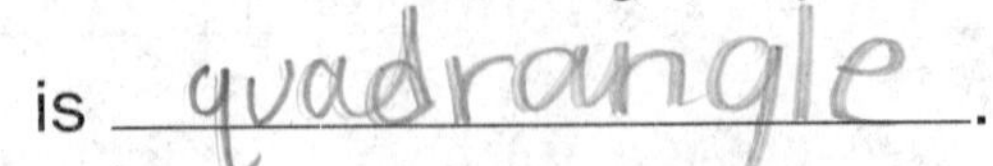

2. Round to the nearest 100 and make an estimate. Then solve. Show your work.

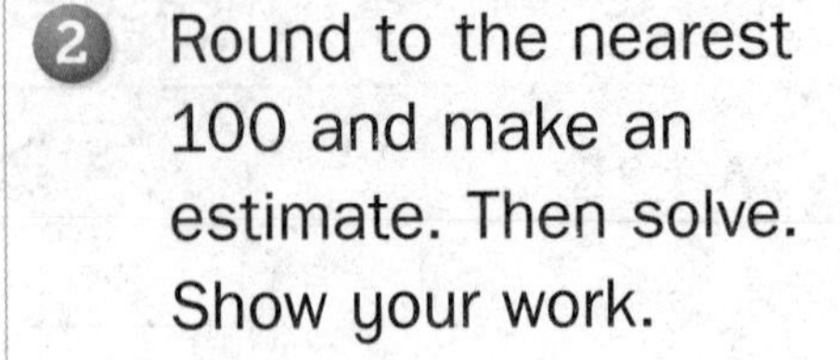

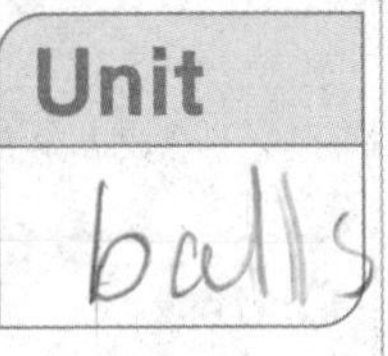

Estimate:

382 − 259 = ______

Think: Does my answer make sense?

3. Draw a sketch to show 20 ÷ 5.

20 ÷ 5 = ______

4. Measure the line segment to the nearest $\frac{1}{2}$ inch and the nearest centimeter.

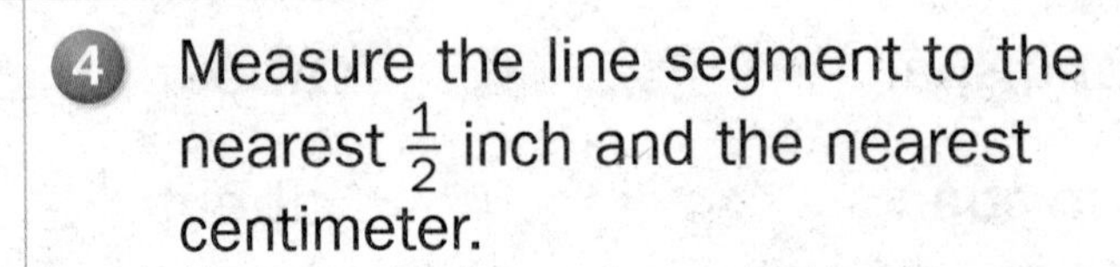

about ______ (unit)

about ______ (unit)

5. **Writing/Reasoning** Explain how you used a tool to measure the line segment to the nearest $\frac{1}{2}$ inch in Problem 4.

Name-Collection Boxes

Lesson 4-9

DATE TIME

1. Three names do not belong in this 100 box. Mark them with an X.

100

980 − 880

25 + 25 + 25

$\begin{array}{r} 80 \\ +\ 30 \\ \hline \end{array}$

30 + 70

$\begin{array}{r} 1,000 \\ -\ 100 \\ \hline \end{array}$

$\begin{array}{r} 63 \\ +\ 37 \\ \hline \end{array}$

2 fifties

$\begin{array}{r} 999 \\ -\ 899 \\ \hline \end{array}$

48 + 52

2. Write at least 10 names for 40.

40

3. Write at least 10 names for 200.

200

4. Write at least 10 names for 1,000.

1,000

Areas of Rectangles

Lesson 4-9

DATE TIME

Math Message

A cloud is partly covering this rectangle. Find the area of the whole rectangle.

Area = 28 square centimeters

Tell a partner how you found the area. Then listen to how your partner found the area. Be ready to share your partner's ideas.

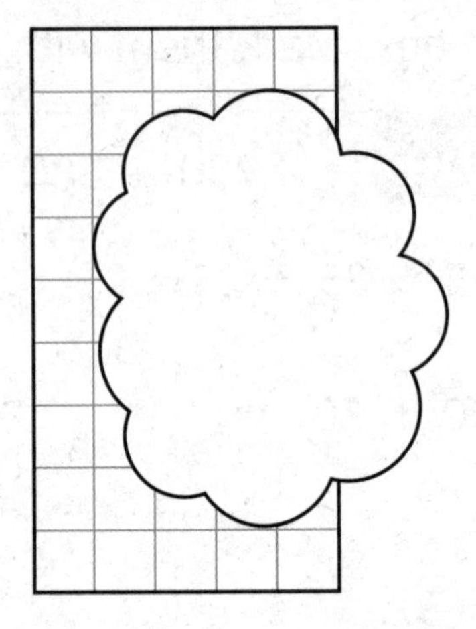

Key:  = square centimeter

---

Listen to your teacher's directions for Problems 1 and 2.

1. Draw a 6-by-4 rectangle.

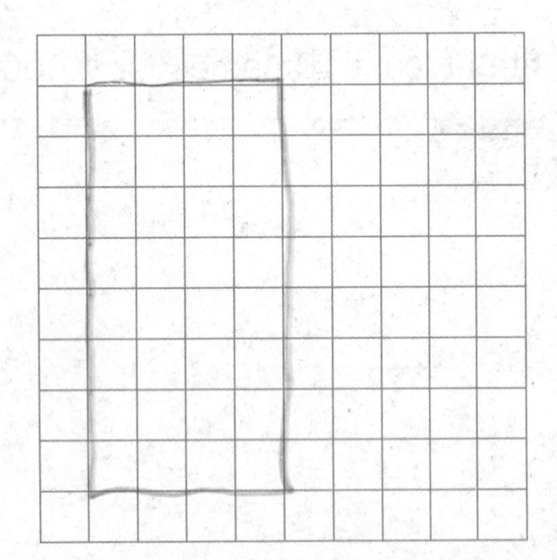

Number of rows: 8

Number of squares in a row: 4

Area = 32 square units

Number sentence: 6 × 4 = 32

2. Draw a 5-by-8 rectangle.

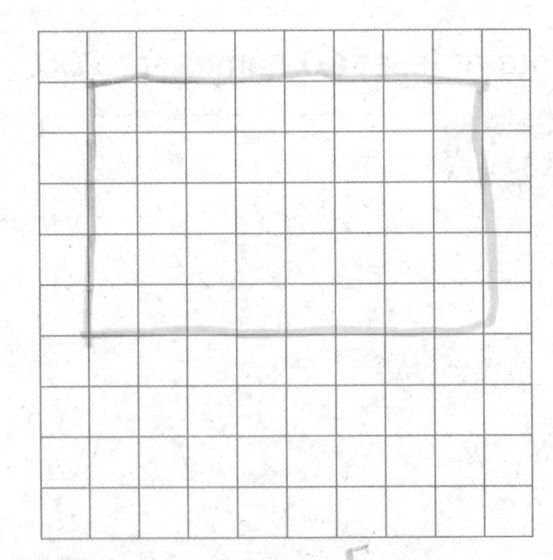

Number of rows: 5

Number of squares in a row: 8

Area = 40 square units

Number sentence: 5 × 8 = 40

Areas of Rectangles (continued)

DATE TIME

Fill in the blanks.

1

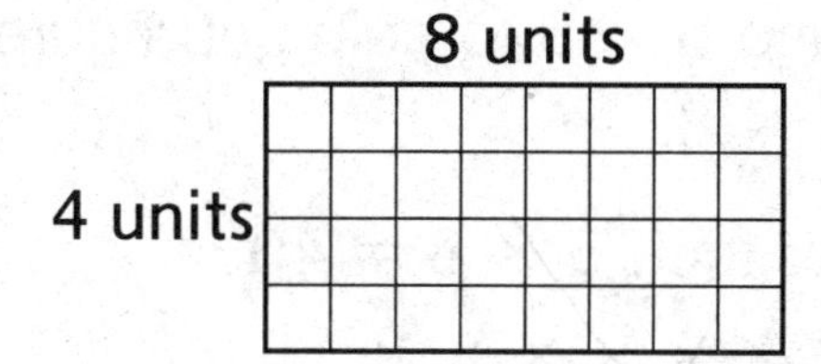

This is a 4-by-8 rectangle.

Area = 32 square units

Number sentence:

4 × 8 = 32

2

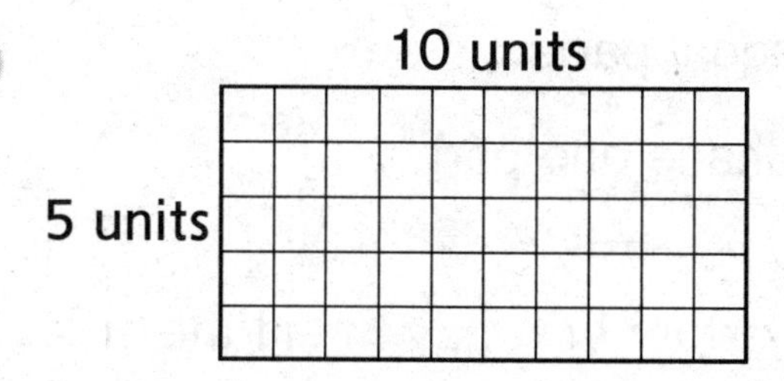

This is a 5-by-10 rectangle.

Area = 50 square units

Number sentence:

5 × 10 = 50

3

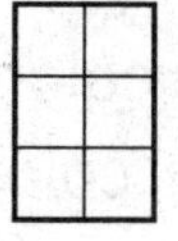

This is a 3-by-2 rectangle.

Area = 6 square units

Number sentence:

3 × 2 = 6

4

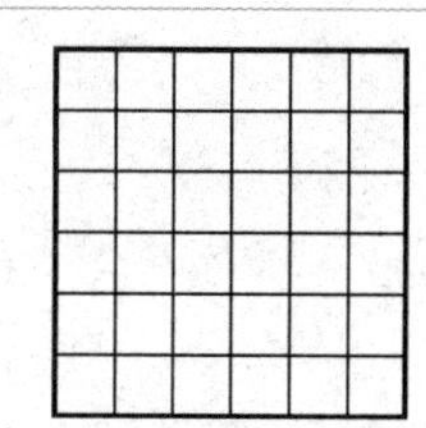

This is a 6-by-6 rectangle.

Area = 36 square units

Number sentence:

6 × 6 = 36

5

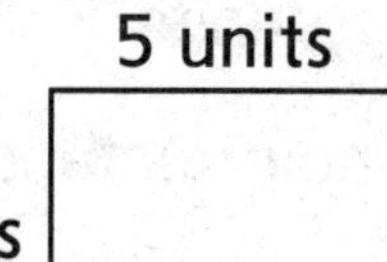

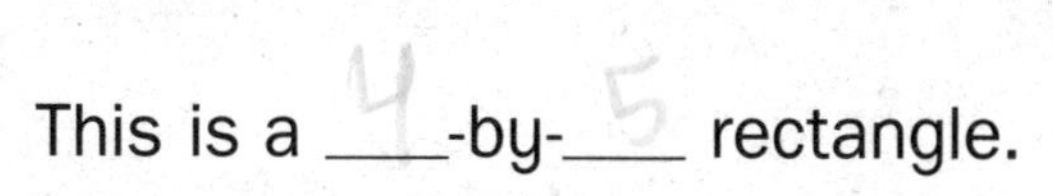

This is a 4-by-5 rectangle.

Area = 20 square units

Number sentence:

4 × 5 = 20

6

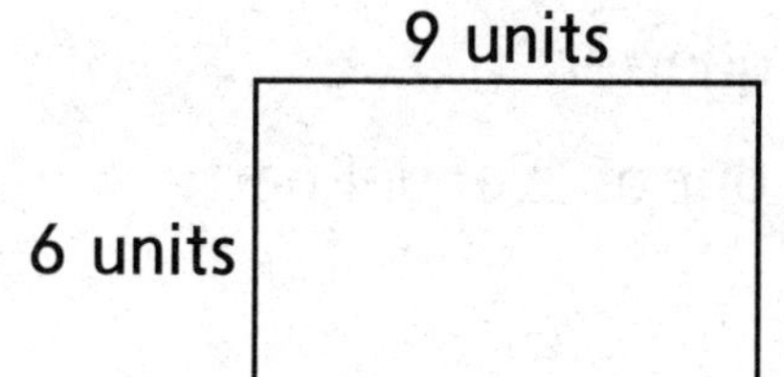

This is a 6-by-9 rectangle.

Area = 54 square units

Number sentence:

6 × 9 = 54

Math Boxes Preview for Unit 5

1 Divide the shape below into four equal parts.

Shade one part.

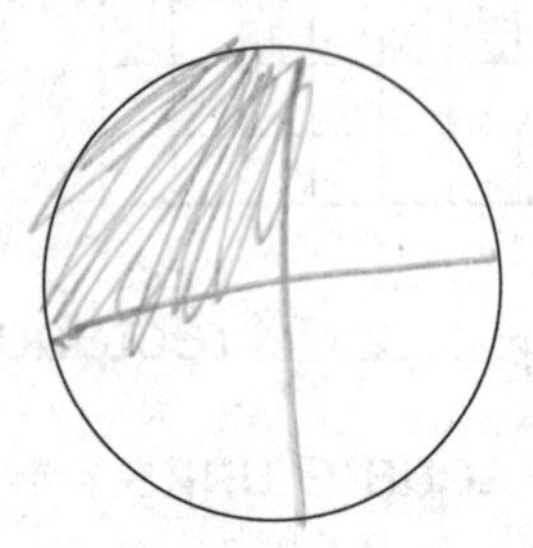

What fraction of the shape is shaded? $\frac{1}{4}$

2 Use the array to show how you can use 5 × 8 to help you figure out 4 × 8.

Helper fact: 5 × 8 = 40

× × × × × × × ×
× × × × × × × ×
× × × × × × × ×
× × × × × × × ×
× × × × × × × ×

4 × 8 = ______

3

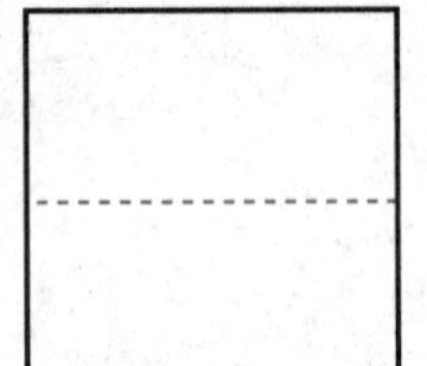

Select all the names that fit one of the parts. Circle the letter(s) next to the correct answer(s).

A. 1-half
B. 1-fourth
C. a quarter
D. 1 out of 2 equal parts

4 Shade the top 2 rows one color and the bottom 2 rows another color.

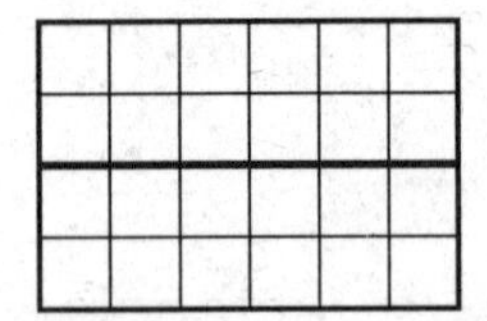

Area of the top 2 rows:
12 square units

Area of the bottom 2 rows:
12 square units

Area of the whole rectangle:
24 square units

5 **Writing/Reasoning** Explain how you found the area of the whole rectangle in Problem 4.

Area and Perimeter

Lesson 4-10

DATE TIME

Math Message

Use the rectangle to answer Problems 1–4.
You may label the side lengths.

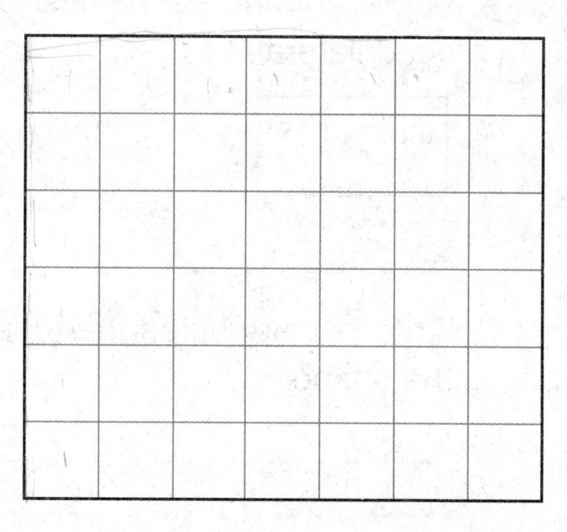

1. Area: 42 square unit (unit)

2. Perimeter: 26 units (unit)

3. Talk to a partner about this rectangle. List all the ways you could find the area.

   6x7=42

   7+7+7+7+7+7 =42

4. List all the ways you could find the perimeter.

   • you can multiply 6x2 and 7x2 and add the products together

   • you can add 6+6 and 7+7 and add the 2 sums.

Math Boxes

Math Boxes

1 Write two different names for this quadrilateral.

parrellogram, quadrangle

Write two attributes that describe the shape.

2 parralel lines

4 sides

SRB 216-217

2 Cross out the 3 names that do NOT belong. Then write the name of the box on the tag.

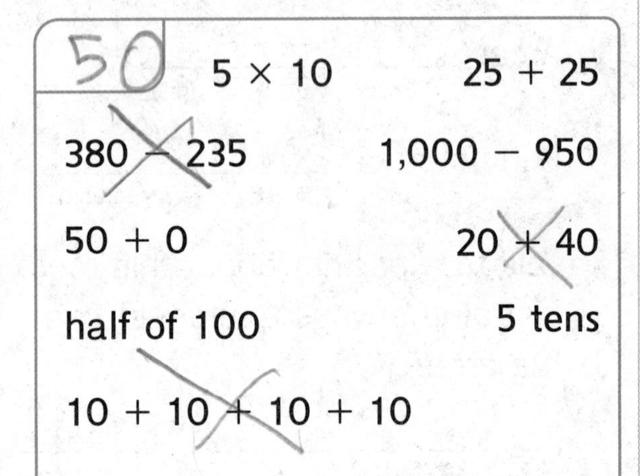

50
5 × 10
25 + 25
380 − 235
1,000 − 950
50 + 0
20 + 40
half of 100
5 tens
10 + 10 + 10 + 10

SRB 96-97

3 Find the perimeter. Fill in the oval next to the correct answer.

2 cm

3 cm

- 5 cm
- 6 cm
- 10 cm
- 12 cm

SRB 174-175

4 Fill in the blanks.

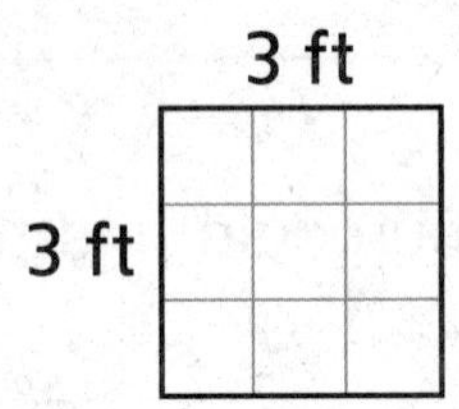

This is a 3-by-3 rectangle.

Area = nine feet (unit)

SRB 176-178

5

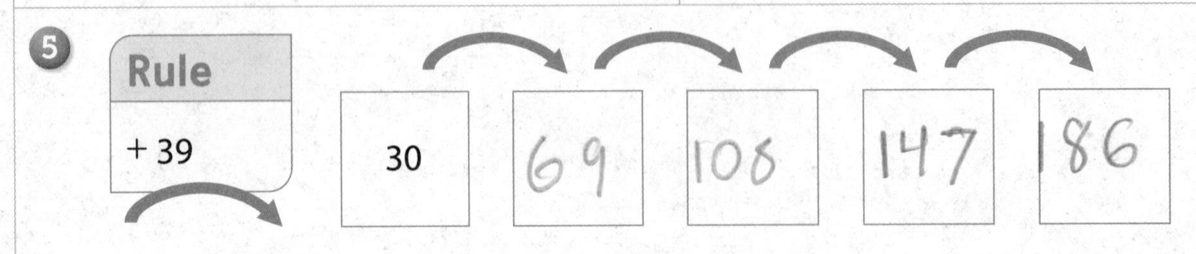

SRB 72-73

Drawing Dog Pens

Lesson 4-11

DATE TIME

Brandi made drawings of her 2 dog pens. She measured the total length of each pen's fence in feet. In her drawings, each square represents 1 square foot.

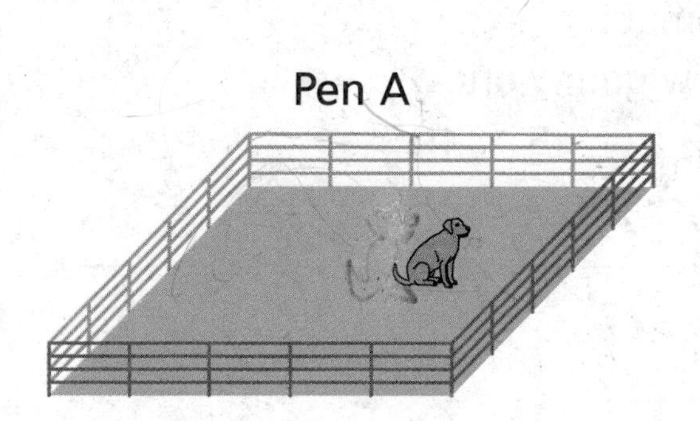

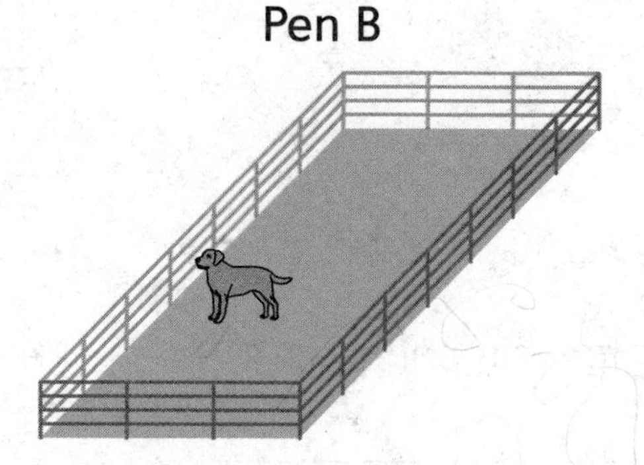

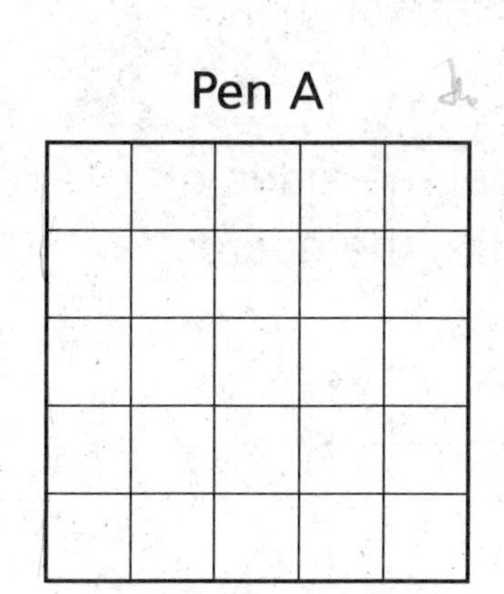

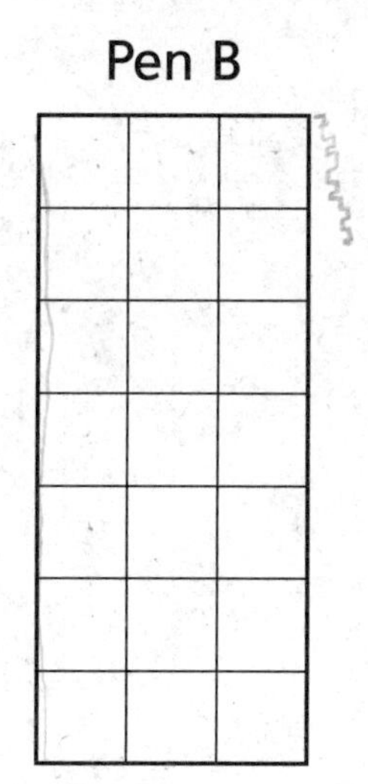

1 ft 1 sq ft

1. Calculate the perimeter and area for each pen. Record the measures using appropriate units.

Pen A

Perimeter = 20 feet

Area = 25 square feet

Pen B

Perimeter = 20 feet (unit)

Area = 21 sq. ft. (unit)

2. What shape are the pens? square

DATE TIME

Math Boxes

1 Draw a quadrilateral that is not a rhombus or a square.

Another name for this shape is a

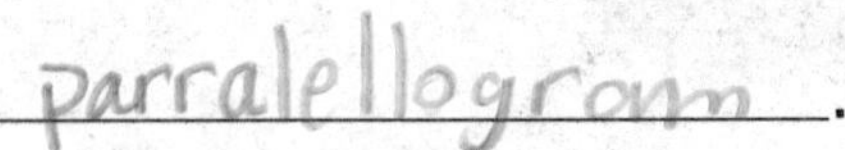.

SRB 216-217

2 Round to the nearest 100 and make an estimate.
Show your work.

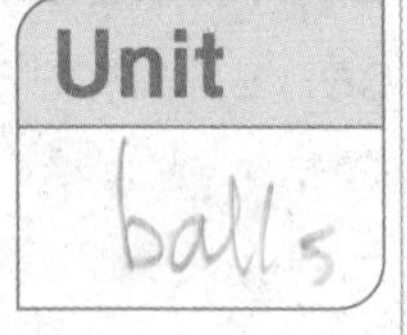

Estimate:

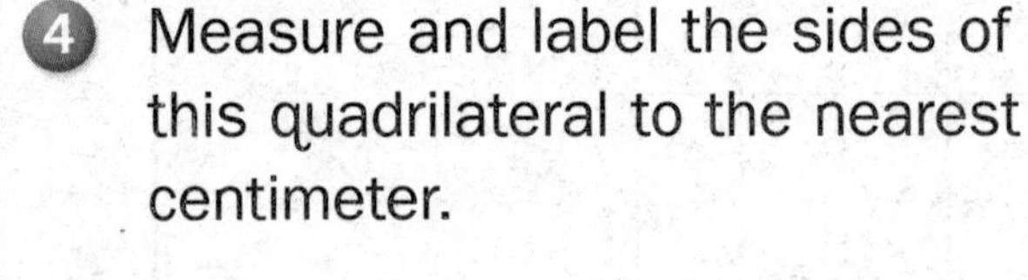

$$\begin{array}{r} 917 \\ -\,283 \\ \hline \end{array}$$

Think: Does my answer make sense?

SRB 106, 119, 122-123

3 Draw a picture to show 25 ÷ 5.

25 ÷ 5 = _______

SRB 39-40, 53

4 Measure and label the sides of this quadrilateral to the nearest centimeter.

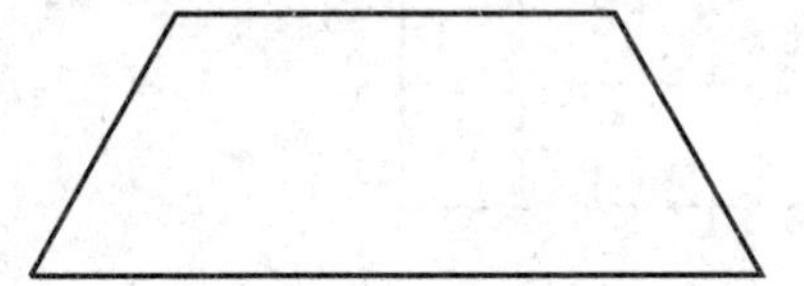

Perimeter: ______________
(unit)

SRB 168-169, 174-175

5 **Writing/Reasoning** Write a number story that fits the number sentence in Problem 3. Include the answer.

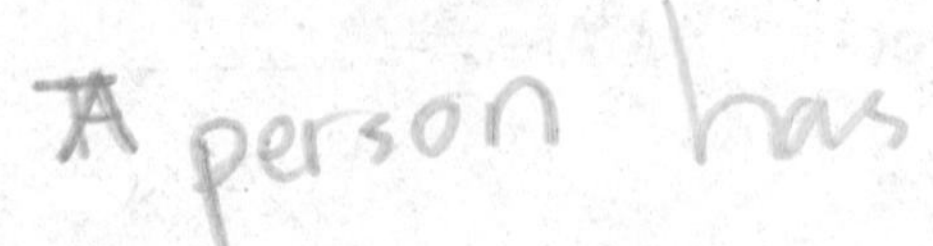

Finding the Areas of Animal Pens

Lesson 4-12

DATE TIME

Decompose each animal pen into rectangles. Write number models for the rectangles and number models for the areas of the pens.

1 Monkeys

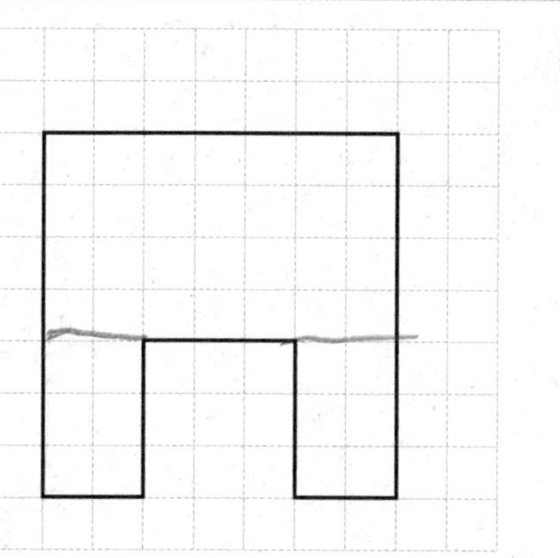

3x2=6 3x2=6 4x7=28

(Number models for rectangles)

28+6+6 = 40

(Number model for area of pen)

Area: 40 square yards

2 Koalas

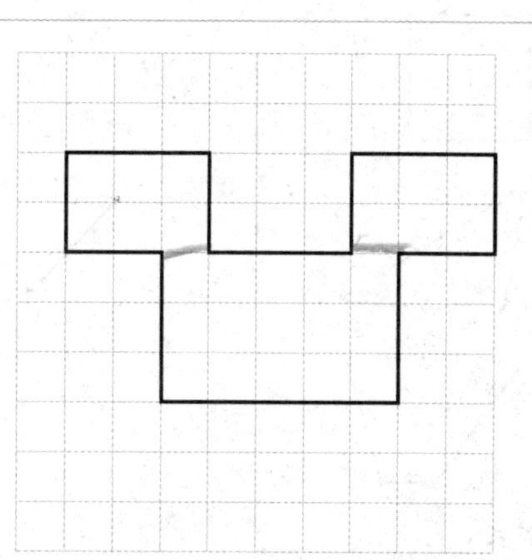

2x3=6 2x3=6 3x5=15

(Number models for rectangles)

15+6+6= 27

(Number model for area of pen)

Area: 27 square yards

3 Prairie dogs

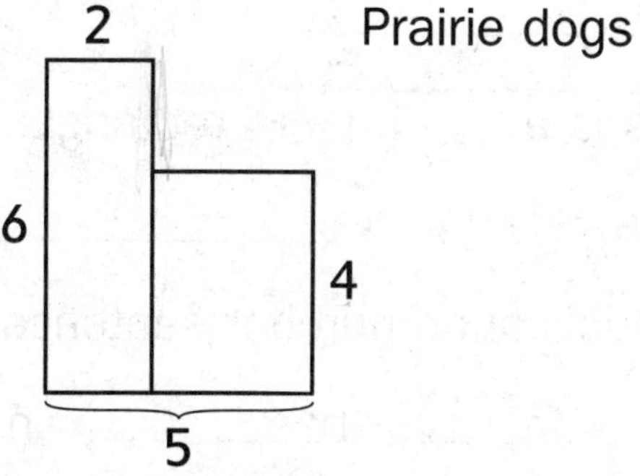

6x2=12 3x4=12

(Number models for rectangles)

12+12=24

(Number model for area of pen)

Area: 24 square yards

4 Giant tortoises

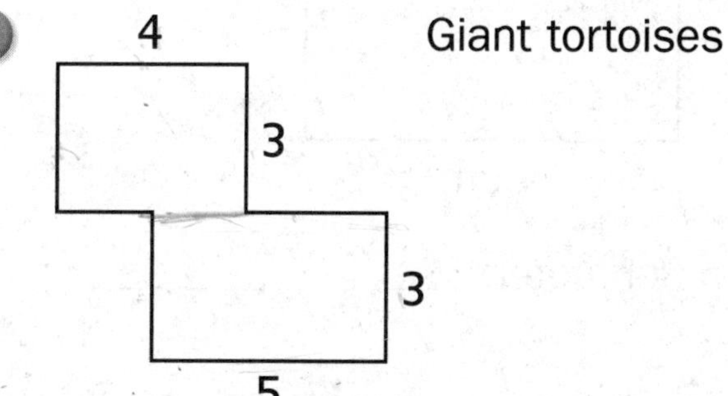

3x4= 12 3x5=15

(Number models for rectangles)

15+12 = 17

(Number model for area of pen)

Area: 17 square yards

Math Boxes

Lesson 4-12

DATE TIME

1. Which shapes are quadrilaterals? Fill in all correct answer(s).

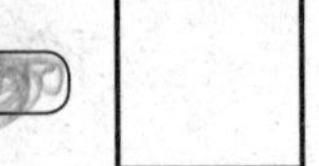

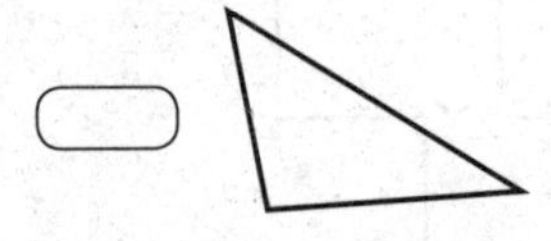
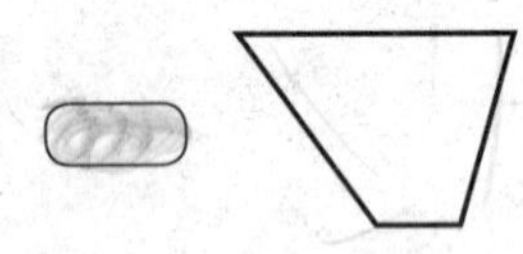
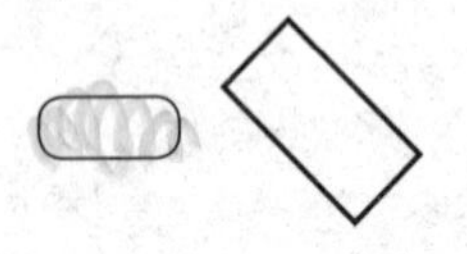

SRB 216-217

2. Write at least 5 equivalent names in the name-collection box.

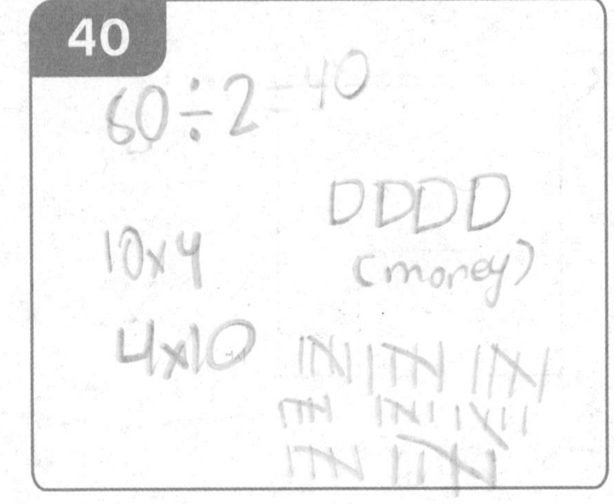

SRB 96-97

3. Find the perimeter of this square.

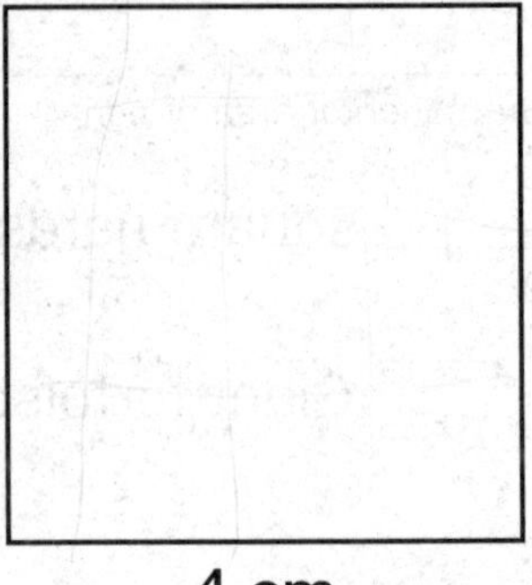

Perimeter: 16cm. (unit)

SRB 174-175

4. Fill in the blanks.

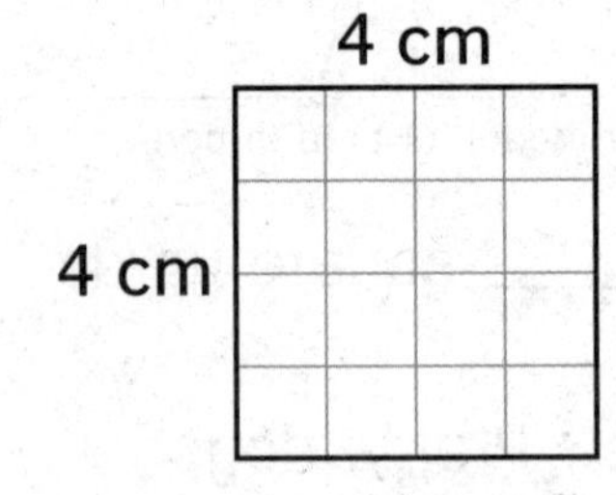

This is a 4-by-4 rectangle.

Area = 16 sq. cm. (unit)

Multiplication number sentence:

4x4=16

SRB 176-179

5.

SRB 72-73

Math Boxes Preview for Unit 5

DATE TIME

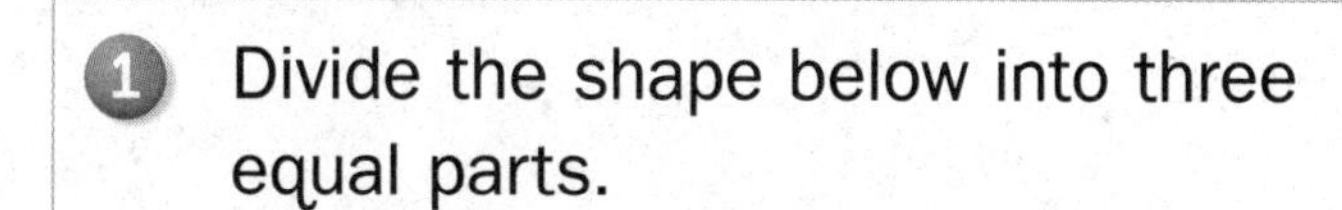

1 Divide the shape below into three equal parts.

Shade two parts.

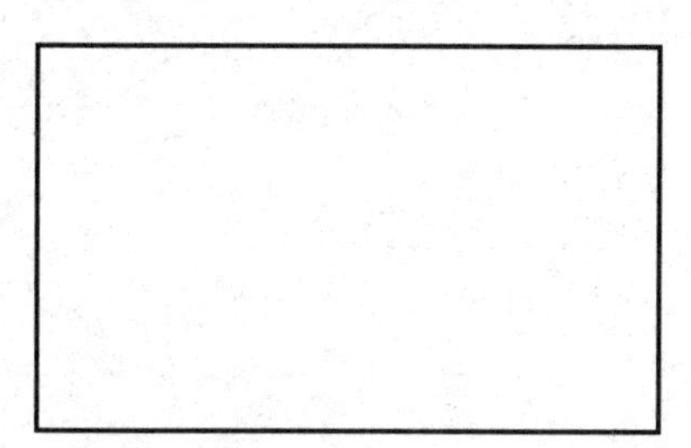

What fraction of the shape is shaded?

____

2 Show how you can use 5×9 to help you figure out 6×9.

× × × × × × × × ×
× × × × × × × × ×
× × × × × × × × ×
× × × × × × × × ×
× × × × × × × × ×

Helper fact: $5 \times 9 = 45$

$6 \times 9 =$ ______

3

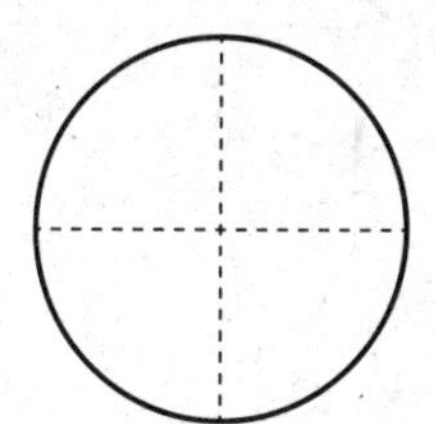

The circle is the whole. What names describe all 4 parts of the whole?

Fill in the circle next to the correct answer(s).

- ◯ **A.** 1-fourth
- ◯ **B.** 4 out of 4 equal parts
- ◯ **C.** 4-fourths
- ◯ **D.** 2 out of 4 equal parts

4 Shade the 3-by-2 rectangle blue.
Shade the 3-by-5 rectangle green.

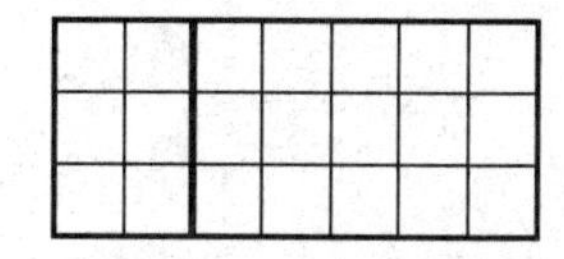

The area of the blue rectangle is ______ square units.

The area of the green rectangle is ______ square units.

The area of the 3-by-7 rectangle is ______ square units.

5 **Writing/Reasoning** Explain how you used the helper fact 5×9 to solve 6×9 in Problem 2.

Notes